Solar Particle Radiation Storms Forecasting and Analysis

Astrophysics and Space Science Library

More information about this series at http://www.springer.com/series/5664

Olga E. Malandraki • Norma B. Crosby
Editors

Solar Particle Radiation Storms Forecasting and Analysis

The HESPERIA HORIZON 2020 Project and Beyond

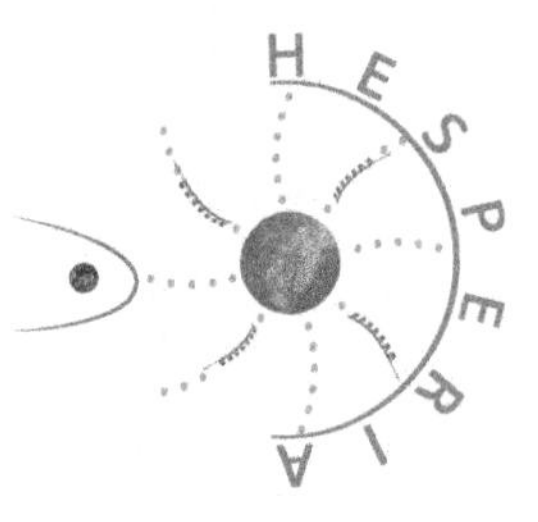

ASSL

Editors
Olga E. Malandraki
National Observatory of Athens
IAASARS
Athens, Greece

Norma B. Crosby
Space Physics Division - Space Weather
Royal Belgian Institute for Space Aeronomy
Brussels, Belgium

ISSN 0067-0057 ISSN 2214-7985 (electronic)
Astrophysics and Space Science Library
ISBN 978-3-319-86766-3 ISBN 978-3-319-60051-2 (eBook)
DOI 10.1007/978-3-319-60051-2

Cover illustration: Space Situational Awareness: Space Weather. Credit: ESA-P. Carril

Printed on acid-free paper

This Springer imprint is published by Springer Nature
The registered company is Springer International Publishing AG
The registered company address is: Gewerbestrasse 11, 6330 Cham, Switzerland

Preface

Ranging in energy from tens of keV to a few GeV solar energetic particles (SEPs) are an important contributor to the characterization of the space environment. Emitted from the Sun they are associated with solar flares and shock waves driven by coronal mass ejections (CMEs). SEP radiation storms may last from a period of hours to days or even weeks and have a large range of energy spectrum profiles. These events pose a threat to modern technology relying on spacecraft and humans in space as they are a serious radiation hazard. Though our understanding of the underlying physics behind the generation mechanism of SEP events and their propagation from the Sun to Earth has improved during the last decades, to be able to successfully predict a SEP event is still not a straightforward process.

The motivation behind the 2-year HESPERIA (High Energy Solar Particle Events forecasting and Analysis) project of the EU HORIZON 2020 programme, successfully completed in April 2017, was indeed to further our scientific understanding and prediction capability of high-energy SEP events by building new forecasting tools while exploiting novel as well as already existing datasets. HESPERIA, led by the National Observatory of Athens, with Project Coordinator Dr. Olga E. Malandraki, was a consortium of nine European teams that also collaborated during the project with a number of institutes and individuals from the international community. The complementary expertise of the teams made it possible to achieve the main objectives of the HESPERIA project:

- To develop two novel real-time SEP forecasting systems based upon proven concepts.
- To develop SEP forecasting tools searching for electromagnetic proxies of the gamma-ray emission in order to predict large SEP events.
- To perform systematic exploitation of novel high-energy gamma-ray observations of the FERMI mission together with in situ SEP measurements near 1 AU.
- To provide for the first time publicly available software to invert neutron monitor observations of relativistic SEPs to physical parameters that can be compared with space-borne measurements at lower energies.

- To perform examination of currently unexploited tools (e.g. radio emission).
- To design recommendations for future SEP forecasting systems.

This book reviews our current understanding of SEP physics and presents the results of the HESPERIA project. In Chap. 1 the book provides a historical overview on how SEPs were discovered back in the 1940s and how our understanding has increased and evolved since then. Current state of the art based on the unique measurements analysed in the three-dimensional heliosphere and the key SEP questions that remain to be answered in view of the future missions Solar Orbiter and Parker Solar Probe that will explore the solar corona and inner heliosphere are also presented. This is followed by an introduction to why SEPs are studied in the first place describing the risks that SEP events pose on technology and human health. Chapters 2 through 6 serve as background material covering solar activity related to SEP events such as solar flares and coronal mass ejections, particle acceleration mechanisms, and transport of particles through the interplanetary medium, Earth's magnetosphere and atmosphere. Furthermore, ground-based neutron monitors are described. The last four chapters of the book are dedicated to and present the main results of the HESPERIA project. This includes the two real-time HESPERIA SEP forecasting tools that were developed, relativistic SEP related gamma-ray and radio data comparison studies, modelling of SEP events associated with gamma-rays and the inversion methodology for neutron monitor observations that infers the release timescales of relativistic SEPs at or near the Sun.

With emphasis on SEP forecasting and data analysis, this book can both serve as a reference book and be used for space physics and space weather courses addressed to graduate and advanced undergraduate students. We hope the reader of this book will find the world of SEP events just as fascinating as we do ourselves.

Olga E. Malandraki
Norma B. Crosby

Acknowledgements

The HESPERIA project has received funding from the European Union's Horizon 2020 research and innovation programme under grant agreement No 637324. The authors thank the EU for this support making it possible to further our knowledge in solar energetic particle research and forecasting, as well as write this book.

The authors of Chaps. 4, 9 and 10 acknowledge the use of ERNE data from the Space Research Laboratory of the University of Turku and of the SEPEM Reference Data Set version 2.00, European Space Agency (2016). They thank the ACE/EPAM, SWEPAM and MAG instrument teams and the ACE Science Center for providing the ACE data. They acknowledge the use of publicly available data products from WIND/SWE and 3DP, GOES13/HEPAD and the CME catalogues from SoHO/LASCO and STEREO/COR1. SoHO is a project of international cooperation between ESA and NASA. They acknowledge also the use of the Harvard-Smithsonian Interplanetary shock Database maintained by M. L. Stevens and J. C. Kasper and of the Heliospheric Shock Database, generated and maintained at the University of Helsinki.

Rolf Bütikofer thanks Erwin Flückiger and Claudine Frieden for their suggestions and assistance in writing Chaps. 5 and 6. This work was supported by the Swiss State Secretariat for Education, Research and Innovation (SERI) under the contract number 15.0233 and by the International Foundation High Altitude Research Stations Jungfraujoch and Gornergrat.

The authors of Chap. 7 thank the National Oceanic Atmospheric Administration (NOAA) for providing GOES data files which were used to calibrate and evaluate the HESPERIA UMASEP-500 tool. They acknowledge the NMDB database (www.nmdb.eu) funded under the European Union's FP7 programme (contract No. 213007). They also acknowledge Dr. Juan Rodriguez from NOAA for his support on the estimation of >500 MeV integral proton flux and expert advice on the GOES/HEPAD data.

The authors of Chap. 8 acknowledge STEREO/HET/LET/SEPT, ACE/EPAM, ACE/SIS, GOES/HEPAD, WIND/3DP and SoHO/ERNE/EPHIN teams as well as the SEPServer team for the availability of the energetic particle data. The STEREO/SEPT and the SoHO/EPHIN projects are supported under grant

50OC1702 by the Federal Ministry of Economics and Technology on the basis of a decision by the German Bundestag. Gerald H. Share and Ronald J. Murphy (Department of Astronomy, University of Maryland, College Park MD 20742 and National Observatory of Athens; Naval Research Laboratory, Washington DC 20375) are acknowledged for making Fermi/LAT data available to the project prior to their publication. Specifically, the authors of Chap. 9 thank G. Share for providing the data on interacting proton spectra derived from the Fermi/LAT γ-ray observations.

Alexandr Afanasiev and Rami Vainio acknowledge the financial support of the Academy of Finland (project 267186) and the computing resources of the Finnish Grid and Cloud Infrastructure maintained by CSC—IT Centre for Science Ltd. (Espoo, Finland) and co-funded by the Academy of Finland and 13 Finnish research institutions. The team of the University of Barcelona has been also partially supported by the Spanish Ministerio de Economía, Industria y Competitividad, under the project AYA2013-42614-P and MDM-2014-0369 of ICCUB (Unidad de Excelencia 'María de Maeztu'). Computational support was provided by the Consorci de Serveis Universtiaris de Catalunya (CSUC).

Alexis P. Rouillard (external collaborator of the HESPERIA project) acknowledges support from the plasma physics data center (Centre de Données de la Physique des Plasmas; CDPP; http://cdpp.eu/), the Virtual Solar Observatory (VSO; http://sdac.virtualsolar.org), the Multi Experiment Data & Operation Center (MEDOC; https://idoc.ias.u-psud.fr/MEDOC), the French space agency (Centre National des Etudes Spatiales; CNES; https://cnes.fr/fr) and the space weather team in Toulouse (Solar-Terrestrial Observations and Modelling Service; STORMS; https://stormsweb.irap.omp.eu/). This includes the data mining tools AMDA (http://amda.cdpp.eu/) and the propagation tool (http://propagationtool.cdpp.eu). He also acknowledges financial support from the HELCATS project under the FP7 EU contract number 606692. The STEREO SECCHI data are produced by a consortium of RAL (UK), NRL (USA), LMSAL (USA), GSFC (USA), MPS (Germany), CSL (Belgium), IOTA (France) and IAS (France).

The authors thank Springer for their interest in the HESPERIA project and the opportunity for the publication of its results.

Contents

List of Abbreviations

ACE	Advanced Composition Explorer
AEPE	Atypical Energetic Particle Event
AIA	Atmospheric Imaging Assembly
AMS	Alpha Magnetic Spectrometer
AU	Astronomical Unit
AWT	Average Warning Time
CDAW	Coordinated Data Analysis Workshops
CGRO	Compton Gamma-Ray Observatory
CME	Coronal Mass Ejection
CORONAS	Complex Orbital near-Earth Observations of Solar Activity
CIR	Corotating Interaction Region
CSA	Coronal Shock Acceleration
CS	Current Sheet
DSA	Diffusive Shock Acceleration
DSP	Downstream Propagation
EGRET	Energetic Gamma Ray Experiment Telescope
EPAM	Electron, Proton, and Alpha Monitor
EPD	Energetic Particle Detector
EPHIN	Electron Proton Helium Instrument
ESA	European Space Agency
ESP	Energetic Storm Particle
FAR	False Alarm Ratio
GBM	Gamma-ray Burst Monitor
GCR	Galactic Cosmic Rays
GLE	Ground Level Enhancement
GME	Goddard Medium Energy
GOES	Geostationary Operational Environmental Satellites
GSE	Geocentric Solar Ecliptic
HCS	Heliospheric Current Sheet
HEPAD	High Energy Proton and Alpha Detector

HESPERIA	High Energy Solar Particle Events forecasting and Analysis
ICME	Interplanetary Coronal Mass Ejection
IMF	Interplanetary Magnetic Field
IMP	International Monitoring Platform
INTEGRAL	INTErnational Gamma-Ray Astrophysics Laboratory
ISIS	Integrated Science Investigation of the Sun
ISS	International Space Station
L1	first Lagrangian point
LAT	Large Area Telescope
LMA	Levenberg-Marquardt algorithm
MED	Medium Energy Detector
MHD	MagnetoHydroDynamics
NASA	National Aeronautics and Space Administration
NGDC	National Geophysical Data Center
NNLS	Non-Negative Least Squares
NOAA	National Oceanic and Atmospheric Administration
NoRP	Nobeyama Radio Polarimeters
NM	Neutron Monitor
PAD	Pitch Angle Distribution
PAMELA	Payload for Antimatter Matter Exploration and Light-nuclei Astrophysics
PFSS	Potential Field Source Surface
POD	Probability of Detection
QLT	Quasi-Linear Theory
PROBA	Project for On-Board Autonomy
REleASE	Relativistic Electron Alert System for Exploration
RHESSI	Ramaty High Energy Solar Spectroscopic Imager
RICH	Ring Imaging CHerenkov
RSTN	Radio Solar Telescope Network
SaP	Shock and Particle
SCR	Solar Cosmic Rays
SDA	Shock Drift Acceleration
SDO	Solar Dynamics Observatory
SEP	Solar Energetic Particle
SEPEM	Solar Energetic Particle Environment Modelling
SMM	Solar Maximum Mission
SOHO	SOlar and Heliospheric Observatory
SolO	Solar Orbiter
SSD	Solid State Detector
STEREO	Solar Terrestrial Relations Observatory
SWAP	Sun Watcher using Active pixel system detector and image Processing

SXR	Soft X-Ray
UMASEP	University of MAlaga Solar particle Event Predictor
WCP	Well-Connected Prediction model
WL	White Light

Chapter 1
Solar Energetic Particles and Space Weather: Science and Applications

Olga E. Malandraki and Norma B. Crosby

Abstract This chapter provides an overview on solar energetic particles (SEPs) and their association to space weather, both from the scientific as well as from the applications perspective. A historical overview is presented on how SEPs were discovered in the 1940s and how our understanding has increased and evolved since then. Current state-of-the-art based on unique measurements obtained in the 3-dimensional heliosphere (e.g. by the Ulysses, ACE, STEREO spacecraft) and their analysis is also presented. Key open questions on SEP research expected to be answered in view of future missions that will explore the solar corona and inner heliosphere are highlighted. This is followed by an introduction to why SEPs are studied, describing the risks that SEP events pose on technology and human health. Mitigation strategies for solar radiation storms as well as examples of current SEP forecasting systems are reviewed, in context of the two novel real-time SEP forecasting tools developed within the EU H2020 HESPERIA project.

1.1 Science

1.1.1 Historical Perspective of Solar Energetic Particle (SEP) Events

It is widely accepted that protons, electrons, and heavier nuclei such as He-Fe are accelerated from a few keV up to GeV energies in at least two distinct locations, namely the solar flare and the coronal mass ejection (CME)-driven interplanetary (IP) shock. The particles observed in IP space and near Earth are commonly referred to as solar energetic particles (SEPs). Those accelerated at flares are known as

O.E. Malandraki (✉)
National Observatory of Athens, IAASARS, Athens, Greece
e-mail: omaland@noa.gr

N.B. Crosby
Royal Belgian Institute for Space Aeronomy, Brussels, Belgium
e-mail: Norma.Crosby@aeronomie.be

O.E. Malandraki, N.B. Crosby (eds.), *Solar Particle Radiation Storms Forecasting and Analysis, The HESPERIA HORIZON 2020 Project and Beyond*, Astrophysics and Space Science Library 444, DOI 10.1007/978-3-319-60051-2_1

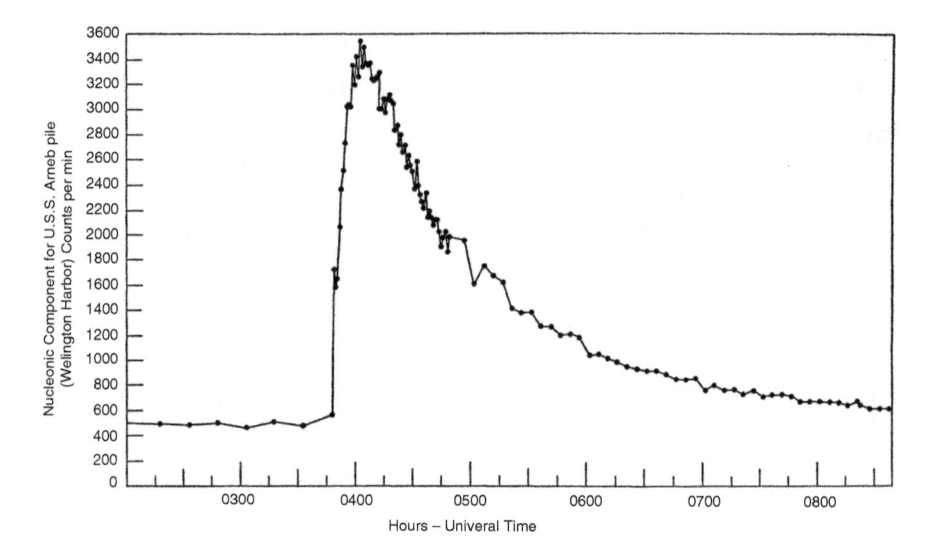

Fig. 1.1 Early observation of a solar energetic particle event (Reproduced from Meyer et al. 1956, permission for reuse from publisher American Physical Society for both print and electronic publication)

impulsive SEP events, particle populations accelerated by near-Sun CME-shocks are termed as gradual SEP events, and those associated with CME shocks observed near Earth are known as energetic storm particle (ESP) events (Desai and Giacalone 2016).

The first SEP event observations extending up to GeV energies were made with ground-based ionization chambers and neutron monitors in the mid 1940s (Forbush 1946). One early event is shown in Fig. 1.1. Until the mid-1990s the so-called 'solar flare myth' scenario was prevalent, in which large solar flares were considered to be the primary cause of major energetic particle events observed at 1 AU (Gosling 1993). However, Wild et al. (1963) had reviewed radio observations and on the basis of the slow-drifting type II bursts observed in close association with the SEP events, proposed an alternative view for the particle acceleration at magnetohydrodynamic shock waves, typically accompanying the flares.

By the end of the 1990s a two-class paradigm of SEP events (see Fig. 1.2) had been generally accepted (e.g. Reames 1999). The flare-related impulsive events lasted a few hours and were typically observed when the observer was magnetically connected to the flare site, were electron-rich and associated with type III radio bursts. These events also had $^{3}He/^{4}He$ ratios enhanced by factors 10^3–10^4, enhanced Fe/O ratios by a factor of 10 over the nominal coronal values, and Fe ionization states of up to $\sim$2. On the other hand, the gradual events lasted several days, had larger fluences, and were attributed to be a result of diffusive acceleration at CME-driven coronal and IP shocks. They were proton-rich, had average Fe/O ratios of 0.1 and Fe ionization states of 14 and were associated with type II radio bursts (e.g. Cliver 2000; Reames 2013).

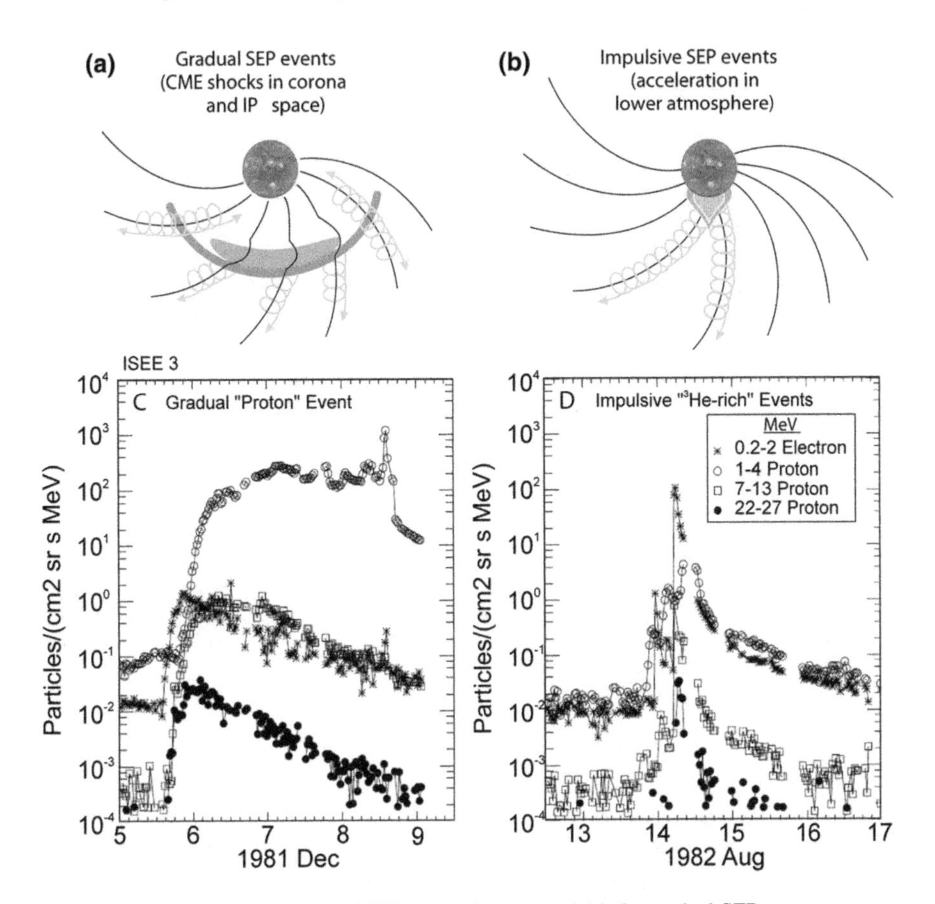

Fig. 1.2 The two-class paradigm of SEP events is presented (**a**) the gradual SEP events occur as a result of diffusive acceleration at CME-driven coronal and IP shocks and populate interplanetary magnetic field (IMF) lines over a large longitudinal extent (**b**) the impulsive SEP events which are produced by solar flares and which populate only those IMF lines well-connected to the flare site. Comparison of intensity-time profiles of electrons and protons in 'pure', (**c**) gradual and (**d**) impulsive SEP events. The gradual event is a disappearing—filament event with a CME but no impulsive flare. The impulsive events come from a series of flares with no CMEs (Reproduced from Desai and Giacalone 2016, permission for reuse from publisher Springer for both print and electronic publication)

Since then, observations have indicated that there are 'hybrid' or mixed event cases, where both mechanisms appear to contribute, with one accelerating mechanism operating in the flare while the other operates at the CME-driven shock (Kallenrode 2003). Such hybrid events may result from the re-acceleration of remnant flare suprathermals by shock waves (Mason et al. 1999; Desai et al. 2006) or from the interaction of CMEs (Gopalswamy et al. 2002). It is noteworthy however, that based on large enhancements in the Fe/O during the initial phases of two large SEP events observed by Wind and Ulysses when the two spacecraft (s/c) were separated by 60° in longitude (Tylka et al. 2013) argued that the initial Fe/O enhancements cannot be cited as evidence for a direct flare component, but instead

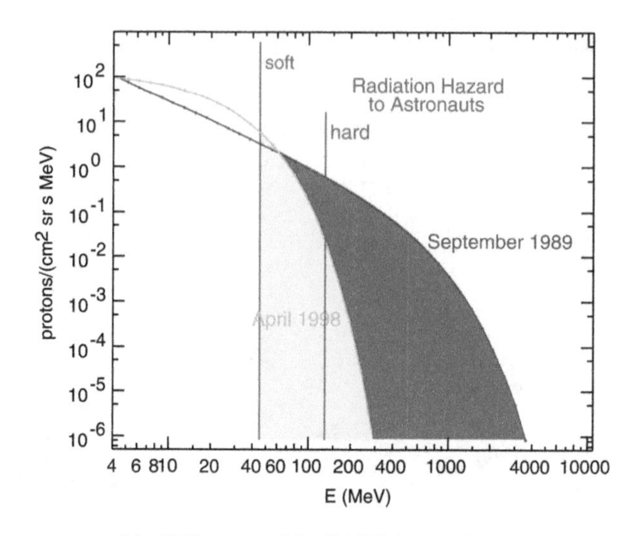

Fig. 1.3 Proton spectra of the SEP events of April 1998 (*green*, Tylka et al. 2000) and September 1989 (*blue*, Lovell et al. 1998) are compared. In *yellow* the hazardous portion of the spectrum during the April 1998 event is highlighted. The region of additional hazardous radiation from the September 1989 event is shaded *red* (Reproduced from Reames 2013, permission for reuse from publisher Springer for both print and electronic publication)

they are better understood as a transport effect, driven by the different mass-to-charge ratios of Fe and O.

High-energy protons in the largest SEP events can pose significant radiation hazards for astronauts and technological systems in space, particularly beyond the Earth's protective magnetic field (National Research Council (NCR) 2008; Cucinotta et al. 2010; Xapsos et al. 2012) (see Sect. 1.2 for more details).

Protons of ~150 MeV are considered as 'hard' radiation since they are very difficult to shield against. Essentially most of the radiation risk of humans in space from SEPs is due to proton intense fluxes of above ~50 MeV, i.e. 'soft' radiation, the energy at which protons begin to penetrate spacesuits and s/c housing. Figure 1.3 compares the proton energy spectra for two large SEP events, presenting typical knee energies of soft and hard radiation SEP events. The most important factor in the radiation dose and depth of penetration of the ions is the location of the energy spectral knee. In yellow in Fig. 1.3, the hazardous part of the spectrum for the April 1998 event is shown, whereas the red shaded area denotes the region of the additional hazardous radiation from the September 1989 event. In the April 1998 event the spectrum rolls over much more steeply at high energies, whereas in the September 1989 event the spectral knee occurred between ~200 and 300 MeV. Events with higher roll-over energies have significantly higher proton intensities above ~100 MeV and can constitute a severe radiation hazard to astronauts (Reames 2013). In fact during the September 1989 event, even an astronaut behind 10 g cm^{-2} of material would receive a dose of ~40 mSv. The annual dose limit for a radiation worker in the United States is 20 mSv (Zeitlin et al. 2013; Kerr 2013). In each solar

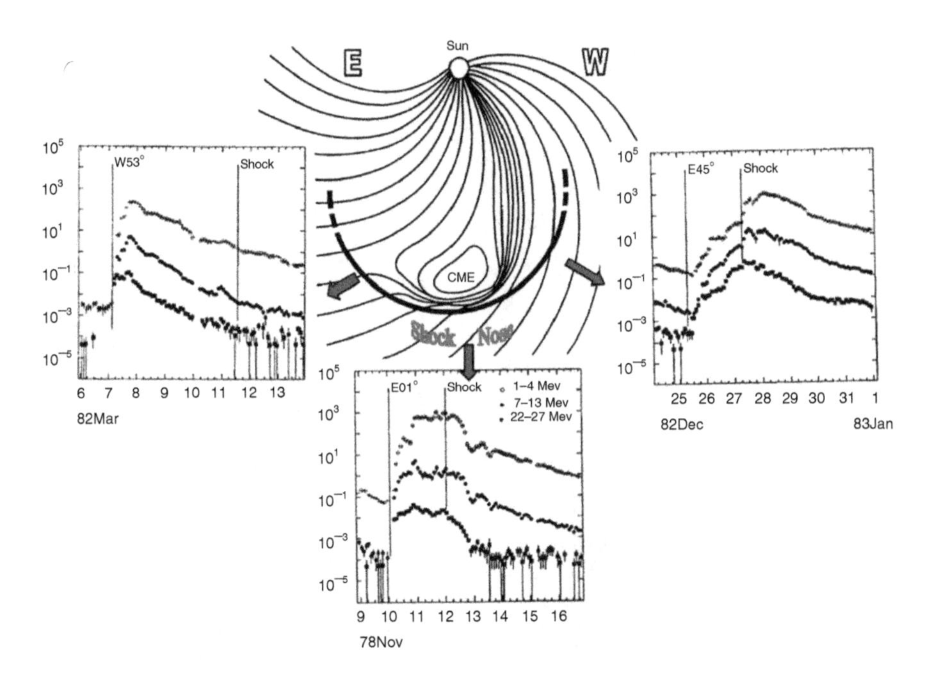

Fig. 1.4 Typical intensity–time profiles of 1–30 MeV protons for gradual SEP events observed at three different solar longitudes relative to the parent solar event. *Dashed lines* indicate the passage of shocks (Reproduced from Reames 2013, permission for reuse from publisher Springer for both print and electronic publication)

cycle several events of this intensity occur, thus, knowledge of the spectral knee energies is essential.

1.1.2 Large Gradual SEP Events

Early multi-spacecraft SEP observations revealed that 1–30 MeV proton time-intensity profiles in large gradual SEP events observed in the ecliptic plane at 1 AU are organized in terms of the longitude of the observer with respect to the traveling CME-driven shock and can be understood if the strongest acceleration occurs near the 'nose' of a CME-driven shock radially expanding outward from the Sun (see Fig. 1.4, Reames 2013; Cane et al. 1988; Cane and Lario 2006).[1] Figure 1.4 shows proton intensity profiles of several SEP events observed by the IMP-8 s/c as a function of longitude of the parent solar event. For observers at solar longitudes to the east of the source, the intensities have rapid rises peaking relatively earlier during the event when there is magnetic connection to the nose of the CME-shock

[1] When observing images of the Sun east and west are reversed.

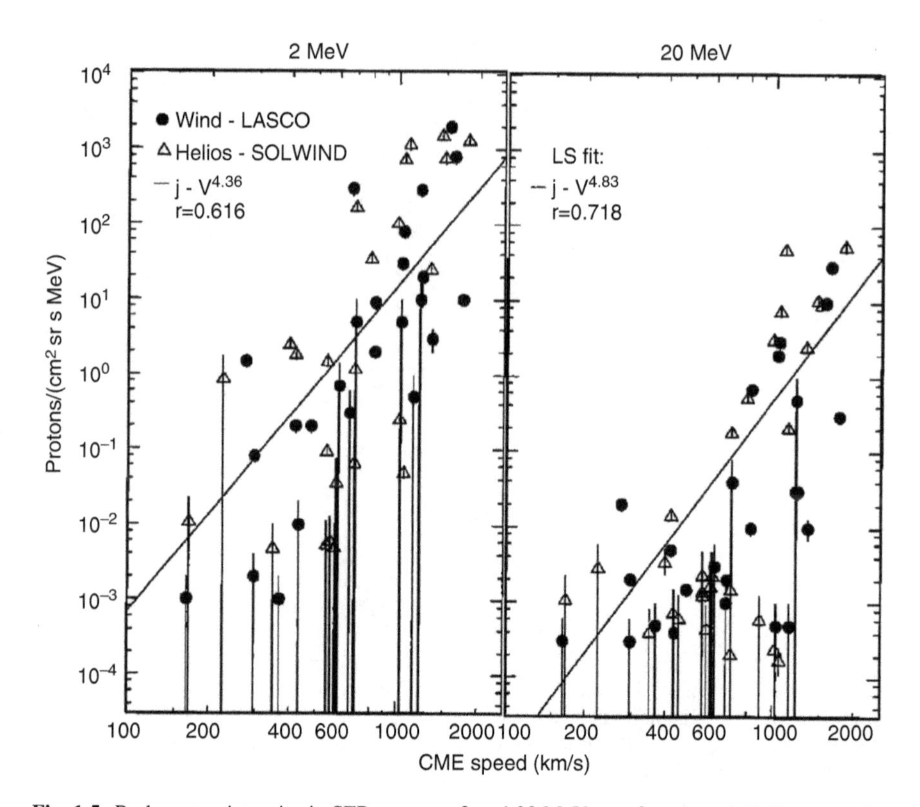

Fig. 1.5 Peak proton intensity in SEP events at 2 and 20 MeV as a function of CME speed. The different symbols denote two combinations of SEP instruments (Wind, Helios) and coronagraphs (LASCO, SOLWIND). Linear least-squares fits as well as the corresponding correlation coefficients are shown for each proton energy (Reproduced from Kahler 2001, permission for reuse from publisher John Wiley and Sons for both print and electronic publication)

near the Sun. Gradual decreasing intensities are observed subsequently, as the shock moves outward and the s/c becomes magnetically connected to the eastern flanks of the shock. For sources near the central meridian the proton intensities peak when the nose of the shock itself arrives at the s/c location. Observers located to the west of the source observe slowly rising intensities that peak after the local passage of the shock.

Comparison between the SEP and the CME or IP shock properties have shown no evidence of a clear correlation. In Fig. 1.5 (Kahler 2001) it is shown that CMEs with similar speeds are associated with a significant spread ($\sim$3–4 orders of magnitude) in the peak proton intensities at 2 and 20 MeV of the associated SEPs at 1 AU. This study subsequently constituted the basis for comparison of a more recent multi-spacecraft study by (Rouillard et al. 2012) in which shock speeds could be measured where the shock intersected the field lines to each s/c in the heliosphere (see Sect. 1.1.4). Using a large number of SEP events, Kahler (2013a) examined the SEP-CME relationship calculating three different SEP event timescales: the onset time from CME launch to the 20 MeV SEP onset time, the rise time from

SEP onset to half the peak intensity and the duration of the SEP intensity above half the peak value. Comparison of these timescales with the CME properties such as speed, acceleration, width and location confirmed that faster (and wider) CMEs drive shocks, and accelerated SEPs over longer periods of time produce SEP events with longer timescales and larger fluences.

A flatter size distribution of SEP events relative to that of flare soft X-ray (SXR) events has been previously reported, with the power-law characterizing SEP size being significantly flatter than that of the SXR flux (e.g. Hudson 1978; Belov et al. 2007; Cliver et al. 2012). Cliver et al. (2012) have shown that this difference is primarily due to the fact that flares associated with large gradual SEP events are an energetic subset of all flares also characteristically accompanied by fast (>1000 km/s) CMEs that drive coronal/IP shock waves. They also concluded that the difference of ~0.15 between the slopes of the SEP event distributions and SEP SXR flares is consistent with the observed variation of SEP event peak flux with SXR peak flux. Kahler (2013b) presented arguments against using scaling laws for the description of the relationship between the size distributions of SXR flares and SEP events. They suggested an alternative explanation for flatter SEP power-law distributions in terms of the recent model of fractal-diffusive self-organized criticality proposed by Aschwanden (2012), providing evidence against a close physical connection of flares with SEP production. Trottet et al. (2015), although based on a limited SEP event sample, have recently studied the statistical relationships between SEP peak intensities of deka-MeV and near-relativistic electrons and characteristic parameters of CME and solar flares: the CME speed as well as the peak flux and fluence of SXR emission and the fluence of microwave emission. Via a partial correlation analysis they showed that the CME speed and SXR fluence are the only parameters that significantly affect the SEP intensity and concluded that both flare acceleration and CME shock acceleration contribute to the deka-MeV proton and near-relativistic electron populations in large SEP events.

Above a few tens of MeV per nucleon, large gradual SEP events are highly variable in their spectral characteristics and elemental composition. As an example, Fig. 1.6 (left) shows the event-integrated Fe/C ratio as a function of energy for the SEP events of April 21, 2002 and August 24, 2002 (Tylka et al. 2005). Both events were associated with flares nearly identical in terms of their sizes and solar locations (~W80), as well as with CMEs with similar speeds of ~2000 km/s, however, there were remarkable differences observed in their associated heavy ion spectral behaviour. To explain these differences, Cane et al. (2003) and Cane et al. (2006) proposed a direct flare particle component above ~10 MeV/nuc and that large SEP events are a mixture of flare-accelerated and shock-accelerated populations. According to this scenario, well-connected western hemisphere events are dominated by flare-accelerated particles above ~10 MeV/nuc, causing the significant increase of Fe/O, and could also account for the increasing energy dependence of the Fe/O ratios observed e.g. during the August 24, 2002 event. On the other hand the CME shock during the April 21, 2002 event is strong enough to accelerate >10 MeV/nucleon particles at 1 AU and lead to the observed Fe/O decrease with increasing energy.

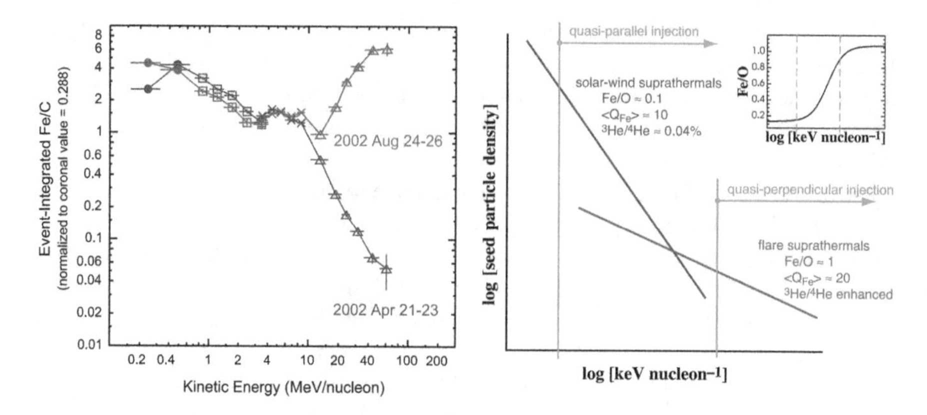

Fig. 1.6 The *left panel* shows a comparison of the energy dependence of the event-integrated Fe/C for the two SEP events of 21 April 2002 (*blue*) and 24 August 2002 (*red*) which are otherwise similar in their properties (Tylka et al. 2005). The *right panel* shows hypothetical spectra of the suprathermal seed populations for shock-accelerated SEPs, comprising both solar wind and flare-accelerated ions. Different injection thresholds will yield different abundance ratios (Reproduced from Reames 2013, permission for reuse from publisher Springer for both print and electronic publication)

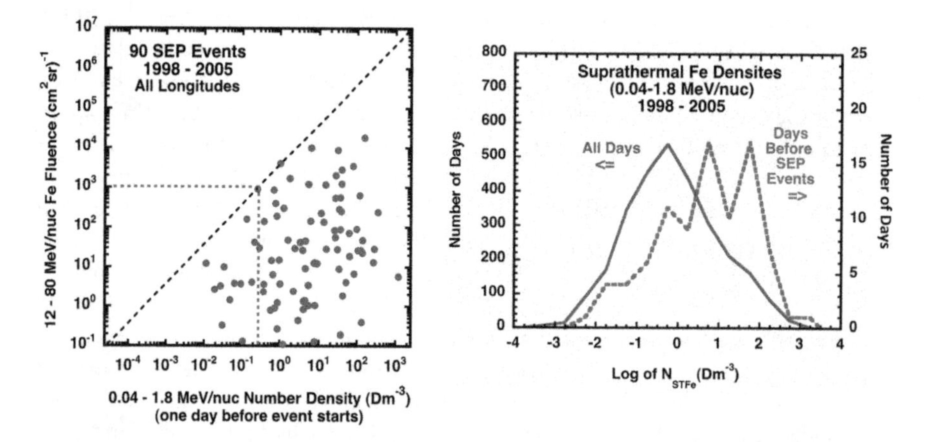

Fig. 1.7 (*left*) The 90 large SEP events defined as events with >12 MeV/nucleon Fe fluences > 0.1/(cm^2 sr) from 1998 to 2005. Days with high fluence only occur when the density of pre-existing suprathermal Fe was >0.3 Dm^{-3}. (*right*) Histogram of daily averaged suprathermal Fe densities for all days from March 1998 to December 2005 (Reproduced from Mewaldt et al. 2012a, permission for reuse from publisher AIP Publishing LLC for both print and electronic publication)

Re-acceleration of remnant flare suprathermals or from accompanying flares has been another plausible idea to account for the observed elemental composition variability in SEP events. Mewaldt et al. (2012a) examined the dependence of SEP fluences on suprathermal seed-particle densities. In Fig. 1.7 (left) the Fe fluence in 90 large SEP events is compared with the pre-existing number density of suprathermal Fe at 1 AU 1 day before the occurrence of the SEP event. They found that the maximum Fe daily-average SEP fluences measured by ACE/SIS are apparently limited by the pre-existing suprathermal number density. In Fig. 1.7 (right) it is shown that the suprathermal Fe densities are significantly greater before the occurrence of these large SEP events with respect to all other days, strongly suggesting that the large fluences of Fe in SEP events only occurred when there was a pre-existing high density of suprathermal Fe. According to these authors remnant flare suprathermal ions, as well as suprathermal material accelerated at previous CME shocks, existed in the heliosphere and served as seed particles subsequently re-accelerated by the CME shock that produced the large CME event (Mason et al. 1999; Desai et al. 2006).

An alternative scenario that (Tylka et al. 2005) proposed is that the observed variability in the energy dependence of the Fe/O ratio could be due to the interplay of two factors namely the evolution in the shock-normal angle as the shock moves outward from the Sun and a compound seed population, typically comprising at least suprathermals from the corona (or solar wind) and flare suprathermals. In this scenario, (Fig. 1.6, right), since the quasi-perpendicular (Q-Perp) shock needs higher injection energy, it may only effectively accelerate impulsive suprathermals originating from the flare acceleration process to high energy, producing the Fe-rich events. On the other hand, since quasi-parallel (Q-Par) shocks have lower injection thresholds they can accelerate the ambient solar wind (or coronal suprathermal ions) producing the Fe-poor events at higher energies. Tylka and Lee (2006) formalized the ideas put forward by Tylka et al. (2005) in an analytical model which above $\sim$1 MeV/nucleon the Tylka and Lee (2006) model reproduced key features of the SEP variability observed in terms of the energy dependence of Fe/O, the ^{3}He/^{4}He ratio and the mean ionic charge state of Fe. Schwadron et al. (2015) further improved the model of coronal shock acceleration. In the left panel of Fig. 1.8, the injection energy of shock-accelerated particles is shown as a function of θ_{Bn} for a range of the perpendicular to parallel diffusion coefficient ratios, whereas in the right panel, the time profiles of the shock or compression radial position (top panel) and θ_{Bn} (bottom

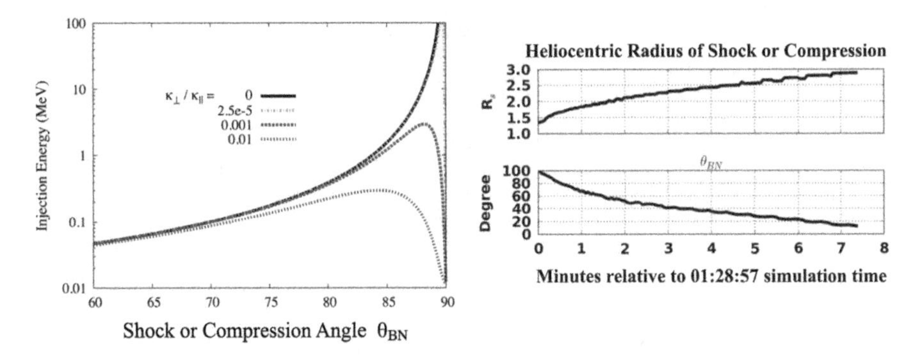

Fig. 1.8 *Left panel*: the injection energy of shock-accelerated particles as a function of θ_{Bn} for a range of the perpendicular and parallel diffusion coefficient ratio. *Right panel*: Time profiles of the shock radial position θ_{Bn} relative to 01:28:57 in the simulation time (see text) (© AAS. Reproduced with permission from Schwadron et al. 2015)

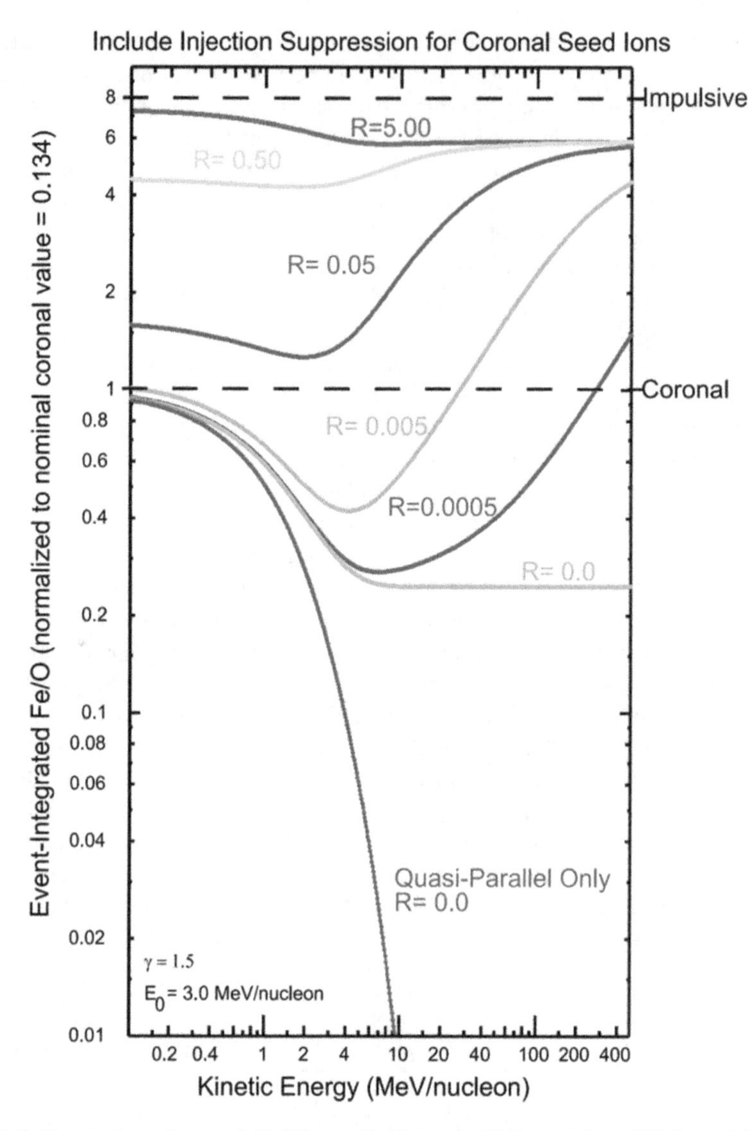

Fig. 1.9 Energy dependence of $(Fe/O)_n$ predicted by the Tylka and Lee (2006) model shown for different values of the parameter R, which reflects the relative strengths of the remnant flare and coronal source contributions at a parallel shock, where seed ions from both populations are injected with equal efficiency (Reproduced from Reames 2013, permission for reuse from publisher Springer for both print and electronic publication)

panel) relative to 01:28:57 in the simulation time are shown. Apparently as the shock moves outward, θ_{Bn} decreases, and the geometry of the shock would change from Q-Perp to Q-Par. Schwadron et al. (2015) noted that the CME expansion and acceleration in the low corona may naturally give rise to rapid particle acceleration and broken power-law distributions in large SEP events. Figure 1.9 shows the results of the (Tylka and Lee 2006) model for the case in which the injection of coronal seed ions at Q-Perp shocks is suppressed. The energy dependence of the normalized Fe/O ratio i.e. $(Fe/O)_n$ is shown for different values of the impulsive suprathermal fraction R in the seed population. In the Q-Par shock event (Fe/O)n $\sim$1 at lower energies ($E < 2$ MeV/nucleon), while $(Fe/O)_n$ monotonically decreases with increasing E. In contrast, in the Q-Perp shock $(Fe/O)_n$ is between 1 and 8 at lower energies, depending on the impulsive suprathermal fractions. With increasing energy the normalized ratio exhibits a complex variation e.g. approaching a plateau or reaching a minimum and further increasing afterwards. Tylka et al. (2005) hence assumed that the high-energy Fe/O ratio could be used as a crude proxy for shock geometry, with Fe-poor and Fe-rich events corresponding to Q-Par and Q-Perp shock geometries, respectively.

It should be noted that these explanations have not taken into account the IP transport effect, which could further distort the Fe/O ratio that emerged from the CME-shock acceleration process (e.g. Tylka et al. 2013). Recently, (Tan et al. 2017) examined 29 large SEP events with peak proton intensity Jpp (>60 MeV) > 1 pfu during solar cycle 23. The emphasis of their examination was put on a joint analysis of the Ne/O and Fe/O data in the 3–40 MeV/nucleon energy range as covered by the Wind/LEMT and ACE/SIS sensors in order to differentiate between the Fe-poor and Fe-rich events at higher energies that emerged from the CME-driven shock acceleration process, after correcting the IP transport effect. One of the main findings of this work is presented in Fig. 1.10 in which the plot of the source plasma temperature T as very recently reported by Reames (2016) versus the normalized Ne/O ratio i.e. $(Ne/O)_n$ at $E = 30$ MeV/nucleon is shown. T is well correlated with $(Ne/O)_n$ with the linear correlation coefficient $(CC) = 0.82$. Therefore, the $(Ne/O)_n$ value at high energies should be a proxy of the injection energy in the shock acceleration process, and hence the shock θ_{Bn} according to the models of Tylka and Lee (2006) and Schwadron et al. (2015).

1.1.3 Ground Level Enhancement (GLE) Events

Ground Level Enhancement (GLE) events form a particular case of high-energy SEP events associated with $\sim$GeV protons. These events pose severe radiation hazards to astronauts and technological assets in space and disrupt airline communications (Shea and Smart 2012). GLEs are nowadays measured with better coverage from space than at ground level, including $\sim$80 MeV/amu to $\sim$3 GeV/amu H and He spectra (Adriani et al. 2011), onsets (Reames 2009a, b), energy spectral shapes and abundances (Mewaldt et al. 2012b), electrons (Kahler 2007, 2012; Tan et al. 2013) and general properties (Gopalswamy et al. 2012). Rouillard et al. (2016) recently studied the link between an expanding coronal shock and the energetic particles measured near Earth during the GLE of 17 May 2012. The analysis showed that the GLE event occurred inside a clear magnetic cloud (see e.g. Malandraki et al. 2002). Using a new technique developed to triangulate the three-dimensional

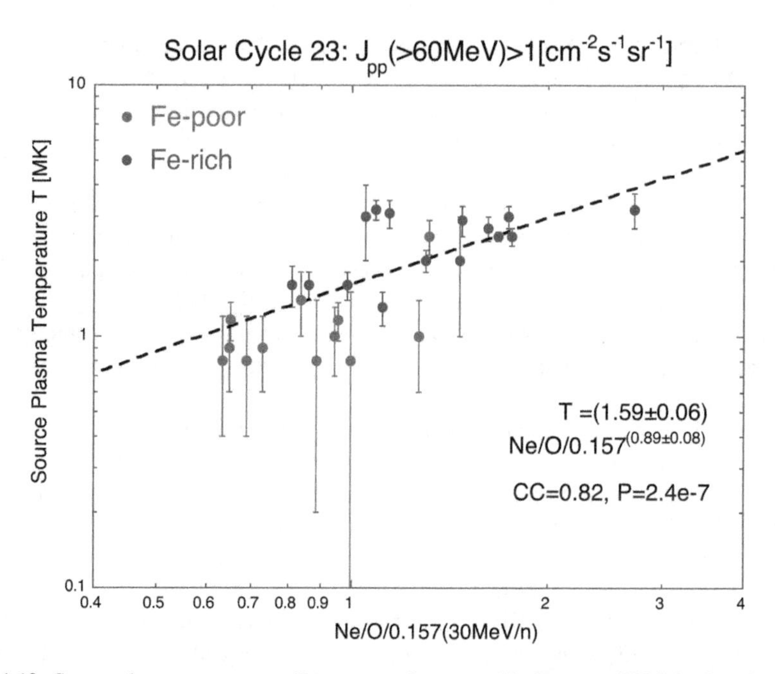

Fig. 1.10 Source plasma temperature T very recently reported by Reames (2016) is plotted vs. the Ne/O/0.157 (30 MeV/nucleon) value for the large Fe-poor (*red*) and Fe-rich (*blue*) events during solar cycle 23. The *dashed line* is the least-square fitting result for all the collected events as listed in Table 2 of Tan et al. (2017) (Reproduced from Tan et al. 2017)

(3D) expansion of the shock forming in the corona it was found that the highest Mach number (M_{FM}) values appear near the coronal neutral line within a few minutes of the CME onset. This neutral line is usually associated with the source of the heliospheric current sheet (HCS) and plasma sheet. It was shown that the release time of GeV particles occurs when the coronal shock becomes super-critical ($M_{FM} > 3$).

1.1.4 Multi-Spacecraft Observations of SEP Events

In this section the differences in the SEP event characteristics as observed from different vantage points in the heliosphere are discussed. Figure 1.11 shows as an example the 1 March 1979 SEP event observed by three different s/c. Helios 1 encounters the event near central meridian and observes a peak in the 3–6 MeV proton intensity near the shock passage time. The intensities at the other s/c, after reaching a peak, begin to track closely those seen at Helios 1 after they enter the so-called 'reservoir' region (see also McKibben 1972; Roelof et al. 1992) in which the intensities and energy spectra are nearly identical. These results indicate that only a small number of particles can leak out of the reservoir. Observations have provided strong evidence for the location of magnetic 'barriers' in space beyond 1 AU and

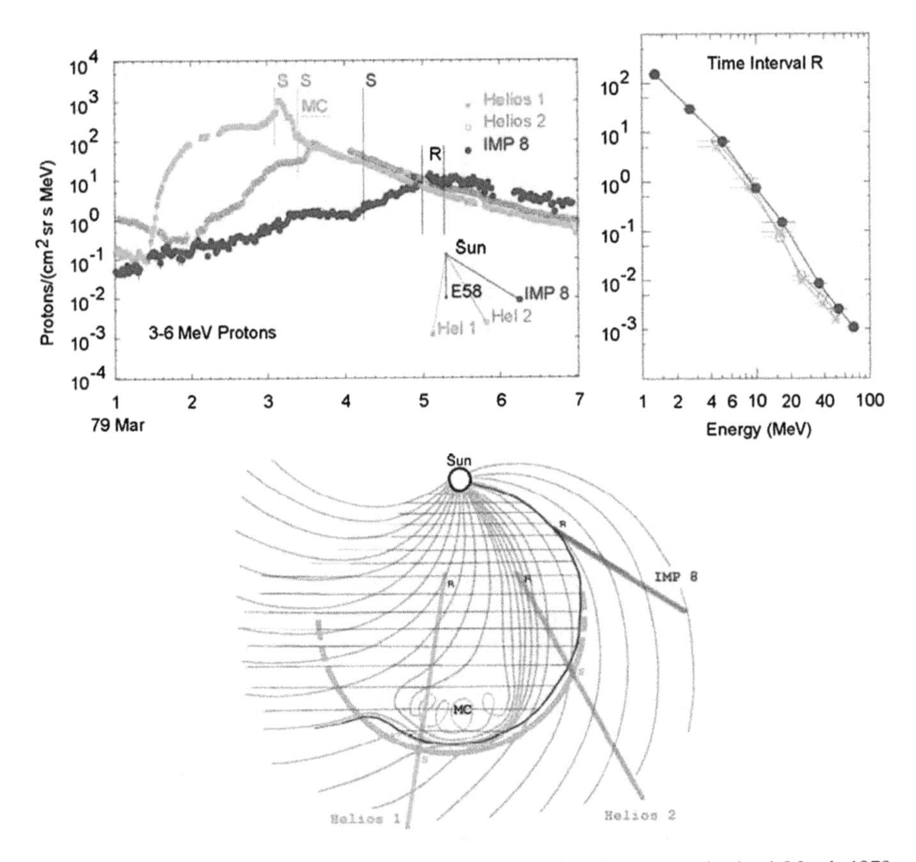

Fig. 1.11 The *top left panel* shows the intensity time-profiles for protons in the 1 March 1979 event at 3 s/c. 'S' denotes the time of shock passage at each s/c. The *top right panel* shows energy spectra in the 'reservoir' region behind the shock at time 'R'. The *lower panel* shows the s/c trajectories through a sketch of the CME (Reproduced from Reames 2013, permission for reuse from publisher Springer for both print and electronic publication)

their role in determining the decay phase of SEP events and the establishment and maintenance of particle reservoirs in the heliosphere (Roelof et al. 1992, 2012a, b; Sarris and Malandraki 2003; Tan et al. 2012). Reames et al. (1996) also considered that the decay phase of the SEP events consists of particles propagating between the converging magnetic field near the Sun and a moving shell of strong scattering formed downstream of the distant traveling shocks. After formation, the reservoir slowly dissipates as a result of the nominal diffusion, convection, adiabatic cooling, and drift mechanisms that govern the propagation of SEPs.

The Ulysses European Space Agency (ESA)/National Aeronautics and Space Administration (NASA) mission provided unprecedented observations of the 3D heliosphere inside ~5 AU. Comparison of simultaneous SEP observations near the ecliptic plane with the Ulysses observations at high latitudes showed that most events that produce large high-energy (>20 MeV) proton and near-relativistic electron flux increases near Earth also produce flux increases at high latitudes, regardless of the longitudinal, latitudinal and radial separation between the s/c, although with somewhat lower maximum intensities and slower rise at Ulysses (McKibben et al. 2003; Lario and Pick 2008; Malandraki et al. 2009). Particle anisotropies during SEP events at high latitudes are typically directed outward from the Sun and aligned with the local magnetic field (McKibben et al. 2003; Malandraki et al. 2009).

The observed field-aligned anisotropies, with components perpendicular to the local magnetic field that are essentially zero, indicate that there is no net flow of particles across the local magnetic field. The Ulysses observations revealed the 3D nature of the reservoir effects in the heliosphere. Dalla et al. (2003) concluded that the presence of a shock is not necessary for creating the near-equality observed at Ulysses and near Earth decay phases, but that these observations are better explained by diffusion across the interplanetary magnetic field (IMF).

More recently, combined observations by the twin STEREO s/c as well as near-Earth observatories revealed the wide longitudinal spreads of large gradual SEP events in the heliosphere and even strongly questioned the constraint of a narrow spread for ^{3}He-rich events (Wiedenbeck et al. 2013). A combination of physical processes appears to cause the large longitudinal spread of high-energy particles. Dresing et al. (2014) concluded that both an extended source region at the Sun and perpendicular transport in the IP medium are involved for most of the wide-spread events under study. The studies (Rouillard et al. 2011, 2012) found that the delayed SEP release times at STEREO and L1 are consistent with either the time required for the CME shock to reach field lines connected to the s/c or with the time required ($\sim$30–40 min) for the CME to perturb the corona over a wide range of longitudes. Observations by Gómez-Herrero et al. (2015) indicated that higher SEP fluxes, harder SEP spectra and direct injections of SEPs onto well-connected IMF lines are associated with lateral expansions of CME-driven shocks in the low corona, and may therefore be responsible for the rapid longitudinal spread as observed at vastly distributed s/c in many SEP events. Other factors that may also play a role in distributing SEP events longitudinally include the large-scale IMF configuration inside interplanetary CMEs (ICMEs) (e.g. Kahler and Vourlidas 2013) and the relative strength of the CME shock, which depends on the local Alfvén speed, rather than the actual speed of the CME (e.g. Gopalswamy et al. 2014).

1.1.5 Particle Acceleration

Recently, important progress has occurred, both from the theoretical and the observational perspective in the research of small-scale magnetic islands in the solar wind and their role in particle acceleration. Khabarova et al. (2015) presented observations that show the occurrence of small-scale magnetic islands and related plasma energization in the vicinity of the HCS. They found evidence that magnetic islands experience dynamical merging in the solar wind and that increases of energetic particle fluxes in the keV–MeV range are found to coincide with the presence of magnetic islands confined by strong current sheets (CSs). Moreover, the interaction of ICMEs with the HCS can lead to significant particle acceleration due to plasma confinement. Their observations confirmed the rippled structure of the HCS and since such a structure confines plasma, it makes possible the strong energization of particles trapped inside small-scale magnetic islands. They concluded that although initial particle acceleration due to magnetic reconnection at the HCS may be insufficient to obtain high energies, the presence of magnetic islands inside the ripples of the HCS or between two CSs with a strong guide field offers the possibility of re-accelerating particles in the ways discussed theoretically by Zank et al. (2014) and le Roux et al. (2015a).

Khabarova et al. (2016) further explored the role of the heliospheric magnetic field configurations and conditions that favor the generation and confinement of small-scale magnetic islands associated with the so-called atypical energetic particle events (AEPEs) in the solar wind. Some AEPEs have been found not to align with standard particle acceleration mechanisms, such as flare-related or simple diffusive shock acceleration processes related to ICMEs and corotating interaction regions (CIRs). They provided more observations fully supporting the idea and the theory of particle energization by small-scale-flux-rope dynamics previously developed by Zank et al. (2014, 2015a, b) and le Roux et al. (2015a, b). If the particles are pre-accelerated to keV energies via classical mechanisms, they may be additionally accelerated up to 1–1.5 MeV inside magnetically confined cavities of various origins. Khabarova et al. (2016) showed that particle acceleration inside magnetic cavities may explain puzzling AEPEs occurring far beyond IP shocks, within ICMEs, before approaching CIRs, as well as between CIRs. SEP transport processes are described in detail in Chap. 4 of this volume (see also Desai and Giacalone (2016) which includes a review on this topic).

1.1.6 Key Open Questions and Future Missions

Solar Orbiter (SolO) is a unique ESA/NASA joint mission conceived to unveil the Sun-heliosphere connection (Mueller et al. 2013), expected to be launched in 2019. The orbital configuration includes a close perihelion, high inclination intervals allowing the observation of the solar polar regions and quasi-co-rotation periods. One of the top-level science questions is "How do solar eruptions produce energetic particle radiation that fills the heliosphere?" which can be broken down into three inter-related key topics: What are the seed populations for energetic particles? How and where are energetic particles accelerated at the Sun? How are energetic particles released from their sources and distributed in space and time?

The Energetic Particle Detector (EPD) instrument suite onboard SolO (Principal Investigator: Prof. J. Rodríguez-Pacheco, Spain) will measure energetic electrons, protons and ions, operating at partly overlapping energy ranges covering from a few keV to 450 MeV/nucleon. The EPD sensors will measure the composition, spectra and anisotropies of energetic particles with sufficient temporal, spectral, angular and mass resolution to achieve the mission goals (Gómez-Herrero et al. 2016). Energetic particles escaping from the acceleration sites propagate through the turbulent IMF. Previous observations by the Helios s/c have shown that SEP events near the Sun are much less affected by IP transport effects compared to 1 AU observations. As the s/c moves further away from the Sun, scattering and diffusion processes become more important and multiple injections closely spaced in time cannot be resolved (Wibberenz and Cane 2006). Thus, SolO observations close to the perihelion will be crucial to unveil SEP injection, acceleration, release and transport processes, in view of the ongoing debate about the SEP acceleration sites, disentangling the acceleration at CME-driven shocks and at reconnection sites in solar flares (e.g. Malandraki et al. 2006).

NASA's Solar Probe Plus mission, recently re-named to Parker Solar Probe to honour pioneering physicist Prof. Eugene Parker, will fly within nine solar radii of the Sun's surface and is scheduled to be launched in July 2018. The two Energetic Particle Instruments (EPI) of the Integrated Science Investigation of the Sun (ISIS) (Principal Investigator: Prof. D. J. McComas, USA) will measure lower (EPI-Lo) and higher (EPI-Hi) energy particles. EPI-Lo will measure ions and ion composition from $\sim$20 keV/nucleon–15 MeV total energy and electrons from $\sim$25–1000 keV. EPI-Hi measures ions from $\sim$1–200 MeV/nucleon and electrons from $\sim$0.5–6 MeV. The unique ISIS observations will allow the exploration of the mechanisms of energetic particles dynamics, including their (1) Origin: defining the seed populations and physical conditions necessary for energetic particle acceleration; (2) Acceleration: determining the role of shocks, reconnection, waves and turbulence in accelerating energetic particles; (3) Transport: revealing how energetic particles propagate from the corona out into the heliosphere (McComas et al. 2016).

It is evident that the next decade is expected to revolutionize our understanding of SEP acceleration and transport, by means of state-of-the-art sensors on board these two upcoming missions providing unique and unprecedented measurements for the exploration of the solar corona and inner heliosphere. Synergies between the two missions are of particular relevance, since both missions have overlapping timelines and the Parker Solar Probe perihelion, reaching up to $\sim$9 solar radii, will permit simultaneous in-situ observations at the SEP acceleration region close to the Sun and at larger radial distances, with continuous remote sensing coverage provided by SolO and near-Earth s/c. Radial alignments between the two s/c will enable the observations of plasma 'entities' from the same solar source region at progressive radial distances as well as the study of energetic particle radial gradients. Furthermore, other useful configurations for the optimization of the science return are alignments along the same IMF line allowing the observation of SEPs originating at the same acceleration site by two or more s/c located at different radial distances. SEP event observations by multiple s/c located at widely separated points in the heliosphere, both in longitude and in latitude, will be valuable for the investigation of the spatial distribution of SEPs and the unraveling of the physical mechanisms responsible for producing wide-spread SEP events (see e.g. Sect. 1.1.4) (Gómez-Herrero et al. 2016).

1.2 Applications

1.2.1 Why Study SEP Events?

It has become apparent during the last decades that SEP events pose important challenges for modern society. Due to their unpredictability, specifically for those that reach relativistic velocities (high energies) and peak values in very short time scales, they are of concern. SEPs ranging from protons to heavy ions up to iron have been found to have impacts on space systems (s/c, instruments, electronic components, solar arrays, . . .), avionics and living organisms (e.g. Feynman and Gabriel 2000; Jiggens et al. 2014). It has even been suggested that systems with very high safety and reliability requirements (e.g. in the nuclear power industry) may need to take account of superstorm ground level radiation on microelectronic devices within the system.

> In the case of nuclear power a Carrington event may not be a sufficient case since relevant timescales for risk assessment may be as long as 10,000 years.
> (Paul Cannon (Cannon et al. 2013))

In the following some of the most important and common SEP induced effects are presented, as well as mitigation strategies currently being relied on.

1.2.2 SEP Effects on Technology

Developments in technology such as miniaturization has no doubt benefited space industry, but at the same time technical equipment has increasingly become more vulnerable to the space environment. On a lesser scale the well-known "snow" effect, resulting from the increase in high energy protons during intense SEP events is sometimes seen on coronagraphic images, as shown in Fig. 1.12, obscuring the image of the CME itself. However, in some instances SEP induced effects may be of such a nature that they can result in long-term damage. Missions that target the inner solar system are especially vulnerable to high-energy charged particles (DiGregorio 2008).

Table 1.1 presents a summary of SEP induced effects observed onboard s/c and aircraft. It is clearly shown how both the energy and species of the particles being considered is an important factor for evaluating their potential effect. Particle flux intensities at lower energies are important for effects such as solar cell degradation, whereas nuclear interactions are associated with particle flux intensities at higher

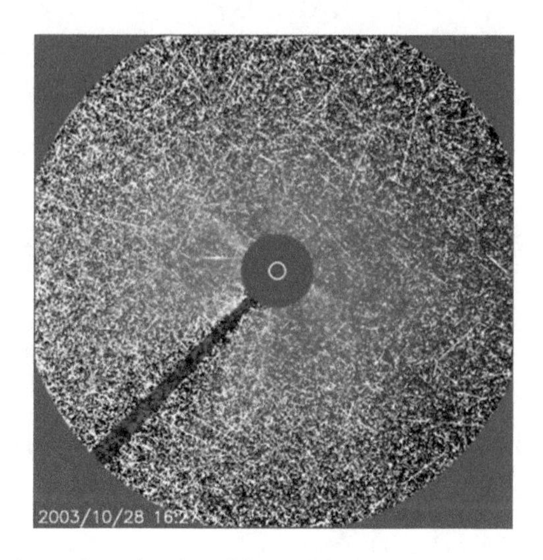

Fig. 1.12 The snowstorm effect observed on the LASCO/SoHO coronagraph on 28 Oct. 2003. Image: ESA/NASA—SOHO/LASCO

Table 1.1 Particle effects on technology observed as a function of the particle energy range

Energy range	Effects
Protons <10 MeV	Material and solar cell effects over time as a result of cumulated dose (e.g. solar cell degradation)
Protons >10 MeV	Nuclear interactions (e.g., sensor background noise, ionization, displacement damage)
Protons >50 MeV and Ions >10 MeV nucleon^{-1}	Nuclear interactions (e.g., single event effects in electronics onboard satellites, as well as aircraft)

energies. Single-event effects (SEEs) are classified as either non-destructive or destructive:

- Single Event Upsets: Occurs in logical circuits and is defined as a bit switching from an initial logical state to an opposite logical state.
- Single Event Latchup: Results in a high operating current, above device specifications, and must be cleared by a power reset.
- Single Event Gate Rupture: Occurs in powerful transistors and is manifested by an increase in gate leakage current
- Single Event Burnout: A condition that can cause device destruction due to a high current state in a power transistor.

For s/c mission planning and operations SEP events are considered. In regard to launch operations the SEP environment is also a decisive factor whether to give the go ahead to launch or not for several reasons. Launch vehicles and s/c reaching sufficiently high geomagnetic latitudes could for example see an increase in SEE rates at times of significant SEP events. On the other hand, optical instruments are also vulnerable to SEPs and induced sensor interference can disrupt the operation of star trackers and put critical s/c manoeuvres at risk.

Under normal space weather conditions Earth's magnetosphere acts as a shield and protects us from charged particles and magnetic clouds. Nevertheless at times SEPs may have sufficient energies to "break" through this shield and enter the ionosphere; SEPs have easier access to the polar regions near Earth's magnetic poles than at the equator due to the "open" magnetic field lines. The cutoff latitude is a function of a particle's momentum per unit charge and is referred to as its rigidity (see Chap. 5). Variations in SEP access to latitudes can occur on time scales of an hour or less in response to changes in the solar wind dynamic pressure and IMF (Kress et al. 2010). For this reason high inclination LEO satellites can at times be vulnerable to SEPs, as well as the International Space Station that has an orbital inclination of 51.64°.

SEP events can also effect signal propagation between Earth and satellites. Polar cap absorption (PCA) events result from intense ionisation of the D-layer of the polar ionosphere by strong (>10 MeV) SEP events. Due mainly to protons with energy in the range of 1–100 MeV (corresponds to an altitude between 30 and 80 km) the increased ionisation absorbs radio waves in the HF and VHF bands, resulting in problems for communications (degraded radio propagation through the polar regions) and navigation position errors with the importance being a function of the individual SEP event.

Despite the relative steepness of SEP energy spectra, the small percentage of protons accelerating up to high energies (>500 MeV) still pose considerable problems. These high-energy SEP events such as the September 1989 SEP event (Fig. 1.3) are often associated with GLE events and can result in secondary radiation caused by particles interacting with s/c shielding and other material. This results in the production of particles such as lower energetic protons, neutrons, and pions that in some cases may be more of an obstacle for the s/c designer than the primary SEPs themselves. While the former can induce SEEs, secondary particle background can have more profound effects on sensitive space-borne instrumentation.

Technology onboard commercial airline operations can also be affected by SEP events including avionics (electronic systems), communications and GPS navigation systems (Jones et al. 2005). Specifically ultra-long-haul "over-the-pole" routes and high-latitude flights are susceptible to these SEP induced effects.

1.2.3 SEPs and Human Health Effects

In addition to being a threat to technology, SEP events are also an important risk to human health. Since the Apollo missions to Earth's Moon in the 1970s human space exploration has mainly been focused on low-Earth orbit altitudes (e.g. Space Shuttle, International Space Station) and suborbital flights. Outside Earth's magnetosphere SEP events have for the most part been a concern for robotic flight missions up until now. During the last decades the vision for space exploration has changed as space agencies and private companies are contemplating sending humans to Mars and asteroids, and as the population on Earth increases colonizing such targets and pursuing deep space exploration will only become more and more attractive. The downside is that human interplanetary exploration will expose astronauts not only to the galactic cosmic ray background but at times also to increased levels of radiation during SEP events and this may indeed be the most important obstacle to overcome.

The field of radiation biology concerns how the radiation environment of space affects cells. Radiation effects on astronauts are sub-classed into two categories:

1. Deterministic (early) Effects: Due to exposure to a large dose of radiation for a limited time (ranges from hair loss, nausea, acute sickness, death)
2. Stochastic (late) Effects: Due to random radiation-induced changes at the deoxyribonucleic acid (DNA) molecule level (cancer).

As already mentioned in Sect. 1.1.1 protons with high energies (>30 MeV) are a health risk for astronauts. For this reason protons with energies >10 MeV are continuously monitored and taken into account when planning extra-vehicular activities.

In those instances SEP events reach aviation altitudes they become also a concern for human health as the radiation dose received can increase. This specifically applies to high-latitude flights (>50°N) and polar routes (>78°N). For commercial aviation this can be a risk for frequent flyers and particularly for aircrew. Effective pilot training programs as well as monitoring, measuring, and regulatory measures in regard to radiation exposure risks for both human tissue and avionics are recognized by a broad community (Tobiska et al. 2015).

1.2.4 Mitigating the Effects of SEPs

The SEP radiation environment is assessed when designing s/c, for s/c mission planning and operations, and when human spaceflight is involved. How does one best go about protecting assets in space and on the ground from the effects of SEP events? For this purpose post-event analysis "hazard assessment" is performed after an anomaly occurs (is recorded). Furthermore, mitigation procedures are put in place before launch (e.g. s/c shielding, redundancy onboard) and during operations SEP forecasting takes place.

1.2.4.1 Hazard Assessment

Analyzing s/c anomalies (hazard assessment) is one way to infer whether an observed anomaly was due to technical or human error, or whether it was a direct consequence of space environment conditions. In Fig. 1.13 it is clearly seen that there was an increase in the number of SEUs in the Ramdisk onboard the LEO Algerian Alsat-1 satellite during three SEP events (29/10/2003: 790 SEUs, 20/01/2005: 774 SEUs, 13/12/2006: 303 SEUs). Figure 1.14 illustrates that the SEU rate is directly a function of the SEP energy spectrum; the flatter the spectrum the higher the number of SEUs (Bentoutou and Bensikaddour 2015).

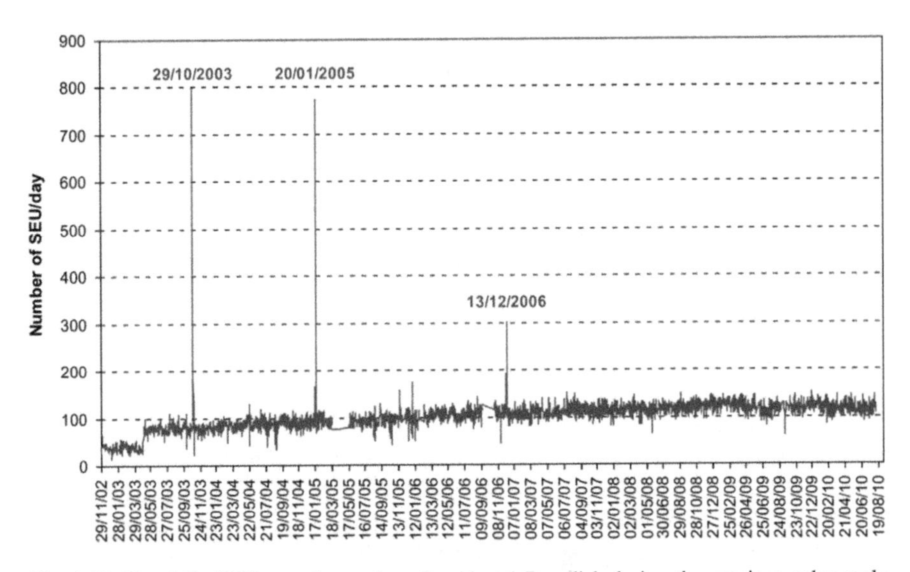

Fig. 1.13 The daily SEU rate observed on the Alsat-1 Ramdisk during the previous solar cycle (Reproduced from Bentoutou and Bensikaddour 2015, permission for reuse from publisher Elsevier for both print and electronic publication)

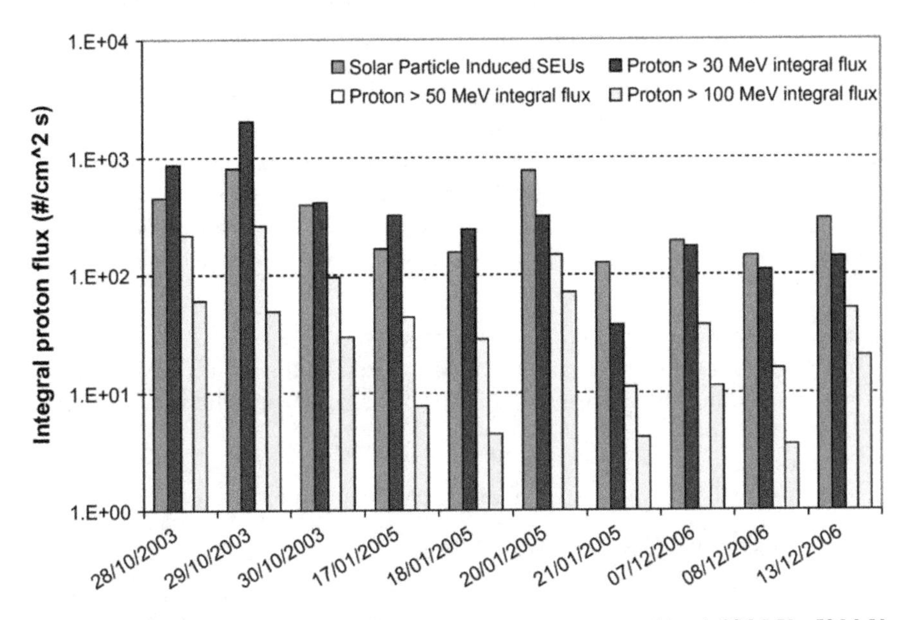

Fig. 1.14 SEU rate (*purple bars*) with proton energy spectrum composition (>30 MeV, >50 MeV, >100 MeV integral flux) when significant SEP events were observed (28–30/10/2003, 17–18/01/2005, 20–21/01/2005, 7–8/12/2006, 13/12/2006) (Reproduced from Bentoutou and Bensikaddour 2015, permission for reuse from publisher Elsevier for both print and electronic publication)

Information obtained from hazard assessment can provide useful input for both engineering (mitigation) and scientific approaches (forecasting), and establish a one-to-one correction between space environment conditions and technical failures. However, s/c operators are sometimes reluctant in providing anomaly data due to confidentiality issues specifically when the anomaly is an important one (e.g. loss of s/c); this makes it sometimes difficult to assess whether a failure was due to the space environment or not. To complement hazard assessment one therefore relies on efficient mitigation strategies such as s/c shielding and forecasting techniques.

1.2.4.2 Mitigation Procedures

The classical engineering approach is based on passive shielding that can protect the crew and hardware (exterior and interior) of the s/c and understanding how the space environment interacts with the shielding. For this reason SEP energy spectra are used as input in engineering tools when computing induced effects such as the dose encountered on technology and humans during SEP events. The shape of the spectrum is important as worst-case scenarios are application dependent, meaning that the flux intensity at lower energies are important for material and solar cell effects. At the other end of the energy spectrum, flux intensities at higher energies are more important for nuclear interactions (e.g., background noise, single event upsets).

Aluminium is generally the material used when building most s/c shielding structures and this type of shielding protects against SEP proton events for the most part (Wilson et al. 1997). However, the higher the particle energy, for example in the case of extreme events, the thicker the shielding material necessary to stop the primary particles. This not only implies the possibility of secondary radiation but also higher costs. For these reasons SEP event forecasting is also relied upon to mitigate against SEP events. Currently, short-term SEP event forecasting systems are based on:

- theoretical understanding (e.g. physical models),
- remote sensing of phenomena such as solar flares, CMEs and active regions
- space-based in-situ observations at L1 (shock arrival, energetic storm particles) and GEO
- historical data
- ground-based observations (e.g. radio, neutron monitors).

and can roughly be divided in two categories:

(a) Physics-based numerical models (e.g. Earth-Moon-Mars Radiation Environment Module (EMMREM) (Schwadron 2010), Predictions of radiation from REleASE, EMMREM and Data Incorporating CRaTER, COSTEP and other SEP measurements (PREDICCS) (Schwadron 2012), Solar Energetic Particle MODel (SEPMOD) (Luhmann et al. 2010), SOLar Particle ENgineering Code (SOLPENCO) (Aran et al. 2006), and SOLPENCO2 (provides SEP modelling away from 1 AU to the SEP statistical model of the SEPEM project (Crosby et al. 2015)))

(b) Empirical models (e.g. University of Malaga Solar Energetic Particle (UMASEP) system (Núñez 2011), Relativistic Electron Alert System for Exploration (REleASE) (Posner 2007), Proton Prediction System (PPS) (Kahler et al. 2007), PROTONS system (Balch 2008), GLE Alert Plus (Kuwabara et al. 2006; Souvatzoglou et al. 2014) and Laurenza's approach (Laurenza et al. 2009))

In some cases forecasting systems rely on methods from both categories such as the SEPForecast tool built under the EU FP7 COMESEP project (263252) (Crosby et al. 2012), (http://www.comesep.eu/alert/).

The EU H2020 HESPERIA project (637324) developed two novel real-time SEP forecasting tools based on the UMASEP and REleASE proven concepts:

- The HESPERIA UMASEP-500 tool makes real-time predictions of the occurrence of GLE events, from the analysis of SXR and differential proton flux measured by the GOES satellite network.
- The HESPERIA REleASE tool generates expected proton flux alerts at two energy ranges (15.8–39.8 MeV and 28.2–50.1 MeV) making use of relativistic electrons ($v > 0.9$ c) provided by the Electron Proton Helium Instrument (EPHIN) on SOHO and near-relativistic ($v < 0.8$ c) electron measurements from the Electron Proton Alpha Monitor (EPAM) aboard the Advanced Composition Explorer (ACE).

Both of these new tools are operational through the project's website (https://www.hesperia.astro.noa.gr/) and described in detail in Chap. 7 of this volume.

Acknowledgements Olga E. Malandraki has been partly supported by the International Space Science Institute (ISSI) in the framework of International Team 504 entitled "Current Sheets, Turbulence, Structures and Particle Acceleration in the Heliosphere".

References

Adriani, O., et al.: Astrophys. J. **742**(2), 102 (2011)
Aran, A.B., et al.: Adv. Space Res. **37**, 1240 (2006)
Aschwanden, M.J.: Astron. Astrophys. **539**, A2 (2012)
Balch, C.C.: Space Weather. **6**, S01001 (2008)
Belov, A., et al.: Solar Phys. **246**(2), 457 (2007)
Bentoutou, Y., Bensikaddour, E.-H.: Adv. Space Sci. **55**, 2820 (2015)
Cane, H.V., Lario, D.: Space Sci. Rev. **123**(1), 45 (2006)
Cane, H.V., et al.: J. Geophys. Res. **93**(A9), 9555 (1988)
Cane, H.V., et al.: Geophys. Res. Lett. **30**, 8017 (2003)
Cane, H.V., et al.: J. Geophys. Res. **111**(A10), A06S90 (2006)
Cannon, P., et al. (eds.): Extreme space weather: impacts on engineered systems and infra-structure. Royal Academy of Engineering, London (2013)

Cliver, E.W.: CP528: acceleration and transport of energetic particles observed in the heliosphere. In: Mewaldt, R.A., et al. (eds.) ACE 2000 Symposium, California, January 2000. AIP Conf. Proc., vol. 528, p. 21. AIP, New York (2000)
Cliver, E.W., et al.: Astrophys. J. Lett. **756**, L29 (2012)
Crosby, N.B., et al.: AIAC '11: space weather: the space radiation environment. Hu, Q., et al. (eds.) 11th Annual International Astrophysics Conference, PalmSprings, USA, March 2012. AIP Conf. Proc., vol. 1500, p. 159. AIP, New York (2012)
Crosby, N., et al.: Space Weather. **13**, 406 (2015)
Cucinotta, F.A., et al.: Space Weather. **8**(12), S00E09 (2010)
Dalla, S., et al.: Geophys. Res. Lett. **30**(1), 8035 (2003)
Desai, M.I., Giacalone, J.: J. Living Rev. Sol. Phys. **13**, 3 (2016)
Desai, M.I., et al.: Astrophys. J. **649**(1), 470 (2006)
DiGregorio, B.E.: Space Weather. **6**, 3 (2008)
Dresing, N., et al.: Astron Astrophys. **567**, A27 (2014)
Feynman, J., Gabriel, S.B.: J. Geophys. Res. **105**(A5), 10543 (2000)
Forbush, S.E.: Phys. Rev. **70**, 771 (1946)
Gómez-Herrero, R., et al.: Astrophys. J. **799**, 55 (2015)
Gómez-Herrero, R., et al.: XXV European Cosmic Ray Symposium, Turin, 4–9 September (2016)
Gopalswamy, N., et al.: Astrophys. J. Lett. **572**(1), L103 (2002)
Gopalswamy, N., et al.: Space Sci. Rev. **171**(1), 23 (2012)
Gopalswamy, N., et al.: Earth Planets Space. **66**, 104 (2014)
Gosling, J.T.: J. Geophys. Res. **98**, 18937 (1993)
Hudson, H.S.: Solar Phys. **57**, 237 (1978)
Jiggens, P., et al.: J. Space Weather Space Clim. **4**, A20 (2014)
Jones, J.B.L., et al.: Adv. Space Res. **36**, 2258 (2005)
Kahler, S.W.: J. Geophys. Res. **106**, 20947 (2001)
Kahler, S.W.: Space Sci. Rev. **129**, 359 (2007)
Kahler, S.W.: Astrophys. J. **769**, 110 (2013a)
Kahler, S.W.: Astrophys. J. **769**, 35 (2013b)
Kahler, S.W., Vourlidas, A.: Astrophys. J. **769**, 143 (2013)
Kahler, S.W., et al.: JASTP. **69**, 43 (2007)
Kahler, S.W., et al.: Space Sci. Rev. **171**, 121 (2012)
Kallenrode, M.B.: J. Phys. G: Nucl. Part. Phys. **29**, 965 (2003)
Kerr, R.A.: Science. **340**(6136), 1031 (2013)
Khabarova, O., et al.: Astrophys. J. **808**, 181 (2015)
Khabarova, O., et al.: Astrophys. J. **827**, 122 (2016)
Kress, B.T., et al.: Space Weather. **8**, 5 (2010)
Kuwabara, T., et al.: Space Weather. **4**, S10001 (2006)
Lario, D., Pick, M.: In: Balogh, A., et al. (eds.) The heliosphere through the solar activity cycle, p. 151. Springer/Praxis Publishing Ltd, Chichester (2008)
Laurenza, M.E., et al.: Space Weather. **7**, 4 (2009)
le Roux, J.A., et al.: Astrophys. J. **801**, 112 (2015a)
le Roux, J.A., et al.: AIAC '14: Linear and nonlinear particle energization throughout the heliosphere and beyond. In: Zank, G.P. (eds). 14th Annual International Astrophysics Conference, Florida, April 2015. J. Phys.: Conc. Ser., vol. 642, p. 012015. IOP Publishing, USA (2015b)
Lovell, J.L., et al.: J. Geophys. Res. **103**, 23733 (1998)
Luhmann, J.G., et al.: Adv. Space Res. **46**, 1 (2010)
Malandraki, O.E., et al.: JASTP. **64**(5–6), 517 (2002)
Malandraki, O.E., et al.: Proceedings of the second solar orbiter workshop, Athens, 16–20 October 2006. ESA Publication (2006)
Malandraki, O.E., et al.: Astrophys. J. **704**, 469 (2009)
Mason, G.M., et al.: Astrophys. J. Lett. **525**, L133 (1999)
McComas, D.J., et al.: Space Sci. Rev. **204**, 187 (2016)
McKibben, R.B.: J. Geophys. Res. **77**(2), 3957 (1972)

McKibben, R.B., et al.: Ann. Geophys. **21**(6), 1217 (2003)
Mewaldt, R.A., et al.: AIAC '11: Space weather: the space radiation environment. Hu, Q. et al. (eds.) 11th Annual International Astrophysics Conference, PalmSprings, USA, March 2012. AIP Conf. Proc., vol. 1500, p. 128. AIP, New York (2012a)
Mewaldt, R.A., et al.: Space Sci. Rev. **171**, 97 (2012b)
Meyer, P., et al.: Phys. Rev. **104**, 768 (1956)
Mueller, D., et al.: Solar Phys. **285**, 25 (2013)
National Research Council (NCR): Managing space radiation risk in the new era of space exploration. The National Academies Press, Washington, DC (2008)
Núñez, M.: Space Weather. **9**, S07003 (2011)
Posner, A.: Space Weather. **5**, S05001 (2007)
Reames, D.V.: Space Sci. Rev. **90**, 413 (1999)
Reames, D.V.: Astrophys. J. **693**, 812 (2009a)
Reames, D.V.: Astrophys. J. **706**, 844 (2009b)
Reames, D.V.: Space Sci. Rev. **175**, 53 (2013)
Reames, D.V.: Sol. Phys. **291**(3), 911 (2016)
Reames, D.V., et al.: Astrophys. J. **466**, 473 (1996)
Roelof, E.C.: AIAC '11: Space weather: the space radiation environment. In: Hu, Q., et al. (eds.) 11th Annual International Astrophysics Conference, PalmSprings, USA, March 2012. AIP Conf. Proc., vol. 1500, p. 174. AIP, New York (2012a)
Roelof, E.C.: AIAC '11: Space weather: the space radiation environment. In: Hu, Q., et al. 11th Annual International Astrophysics Conference, PalmSprings, USA, March 2012. AIP Conf. Proc., vol. 1500, p. 180. AIP, New York (2012b)
Roelof, E.C., et al.: Geophys. Res. Lett. **19**, 1243 (1992)
Rouillard, A., et al.: Astrophys. J. **735**, 7 (2011)
Rouillard, A.P., et al.: Astrophys. J. **752**, 44 (2012)
Rouillard, A.P., et al.: Astrophys. J. **833**, 45 (2016)
Sarris, E.T., Malandraki, O.E.: Geophys. Res. Lett. **30**(21), 2079 (2003)
Schwadron, N.A.: Space Weather. **8**, 1 (2010)
Schwadron, N.A.: Space Weather. **10**, 10 (2012)
Schwadron, N.A., et al.: Astrophys. J. **810**(2), 97 (2015)
Shea, M.A., Smart, D.F.: Space Sci. Rev. **171**, 161 (2012)
Souvatzoglou, G., et al.: Space Weather. **12**, 11 (2014)
Tan, L.C., et al.: Astrophys. J. **750**, 146 (2012)
Tan, L.C., et al.: Astrophys. J. **768**, 68 (2013)
Tan, L.C., et al.: Astrophys. J. **835**(2), 192 (2017)
Tobiska, W.K., et al.: Space Weather. **13**, 4 (2015)
Trottet, G., et al.: Solar Phys. **290**, 819 (2015)
Tylka, A.J., Lee, M.A.: Astrophys. J. **646**(2), 1319 (2006)
Tylka, A.J., et al.: CP528: Acceleration and transport of energetic particles observed in the heliosphere. In: Mewaldt, R.A., et al. (eds.) ACE 2000 Symposium, California, January 2000. AIP Conf. Proc., vol. 528, p. 147. AIP, New York (2000)
Tylka, A.J., et al.: Astrophys. J. **625**, 474 (2005)
Tylka, A.J., et al.: Sol. Phys. **285**, 251 (2013)
Wibberenz, G., Cane, H.V.: Astrophys. J. **650**, 1199 (2006)
Wiedenbeck, M.E., et al.: Astrophys. J. **762**, 54 (2013)
Wild, J.P., et al.: Annu. Rev. Astron. Astrophys. **1**, 291 (1963)
Wilson, J.W., et al. (ed.): Shielding strategies for human space exploration, NASA Conference Publication, 3360 (1997)
Xapsos, M.A., et al.: IEEE Trans. Nucl. Sci. **59**(4), 1054 (2012)
Zank, G.P., et al.: Astrophys. J. **797**, 28 (2014)

Zank, G.P., et al.: Astrophys. J. **814**, 137 (2015a)
Zank, G.P., et al.: AIAC '14: linear and nonlinear particle energization throughout the heliosphere and beyond. In: Zank, G.P. (ed.) 14th Annual International Astrophysics Conference, Florida, April 2015. J. Phys.: Conc. Ser., vol. 642, p. 012031. IOP Publishing, USA (2015b)
Zeitlin, C., et al.: Science. **340**(6136), 1080 (2013)

Chapter 2
Eruptive Activity Related to Solar Energetic Particle Events

Karl-Ludwig Klein

Abstract Solar energetic particle events are associated with solar activity, especially flares and coronal mass ejections (CMEs). In this chapter a basic introduction is presented to the nature of flares and CMEs. Since both are manifestations of evolving magnetic fields in the solar corona, the chapter starts with a qualitative description of the magnetic structuring and electrodynamic coupling of the solar atmosphere. Flares and the radiative manifestations of energetic particles, i.e. bremsstrahlung, gyrosynchrotron and collective plasma emission of electrons, and nuclear gamma-ray emission are briefly presented. Observational evidence on the particle acceleration region in flares is given, as well as a very elementary qualitative overview of acceleration processes. Then CMEs, their origin and their association with shock waves are discussed, and related particle acceleration processes are briefly described.

2.1 Introduction

Solar energetic particle events are associated with transient solar activity, especially with flares and coronal mass ejections (CMEs). The understanding of how and when the sun ejects enhanced fluxes of protons, ions and electrons, sometimes up to relativistic energies, needs insight into these basic eruptive processes. In this chapter an elementary introduction is presented. Particle acceleration requires transient electric fields. They are produced in relation with magnetic reconnection and turbulence, and in large-scale shock waves driven by CMEs. Since flares and CMEs often occur together, it is not easy to identify which of the candidate acceleration processes is at work. They may all act together, but provide particles of different energies.

K.-L. Klein (✉)
LESIA-Observatoire de Paris, CNRS, 92190 Meudon, France

PSL Research University, Universités P & M. Curie, Paris-Diderot, Meudon, France
e-mail: ludwig.klein@obspm.fr

O.E. Malandraki, N.B. Crosby (eds.), *Solar Particle Radiation Storms Forecasting and Analysis, The HESPERIA HORIZON 2020 Project and Beyond*, Astrophysics and Space Science Library 444, DOI 10.1007/978-3-319-60051-2_2

The chapter starts with a brief overview of the magnetic structuring of the outer solar atmosphere (Sect. 2.2). Flares and accelerated particle signatures related to them in the solar atmosphere are introduced in Sect. 2.3, together with the radiative processes that make these particles observable at gamma-ray, hard X-ray and radio wavelengths. CMEs, shock waves and the related particle acceleration processes are addressed in Sect. 2.4. Because of the introductory nature of this chapter, references are rather to review papers than to the original literature.

2.2 The Scene

The solar corona is a hot plasma with an average ion temperature $T \simeq 1.5 \cdot 10^6$ K. The mean energy of the particles in this plasma, to the extent that it can be described by a Maxwellian distribution, is $kT \simeq 160$ eV, where k is Boltzmann's constant.

A remarkable and significant feature revealed by eclipse observations (Fig. 2.1a) is the non-spherical shape of the corona. In an eclipse image one sees especially light from the photosphere, which is Thomson-scattered by free electrons in the corona. The morphology of the corona hence shows the electron density integrated along the line of sight. The image demonstrates that gravity is not the only force that comes into play. The electron concentrations shown by the bright localized structures are confined by magnetic fields. This is also shown by the EUV multi-wavelength image from the *Solar Dynamics Observatory* in Fig. 2.1b: this emission is a mixture of bremsstrahlung and spectral lines emitted mostly at MK-temperatures. The bright structures (active regions) visualize coronal magnetic field lines connected to the underlying atmosphere and the solar interior. They overlie regions with strong magnetic fields in the photosphere, often including sunspots.

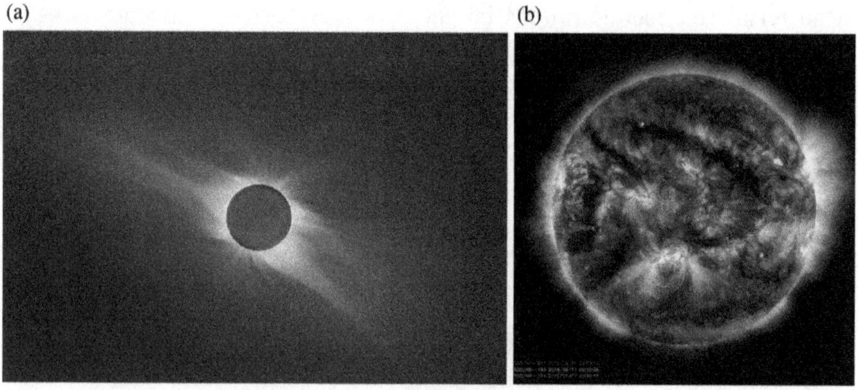

(a) (b)

Fig. 2.1 The solar corona during a total eclipse in visible light (**a**; courtesy C. Viladrich) and as seen in EUV (**b**; courtesy of NASA/SDO and the AIA, EVE, and HMI science teams)

The localized magnetic fields in active regions are the emerged parts of a global solar magnetic field. The plume-like structures pointing to the lower left and upper right in the eclipse image show clearly that such a global magnetic field exists. In and below the photosphere this field is subject to the motions of the plasma, such as the convective motions revealed by granulation and super granulation in the photosphere. These motions shuffle field lines around in the photosphere, concentrate them into magnetic flux tubes with a strong magnetic field that is surrounded by regions with no or weak magnetic field. As the kinetic and dynamic pressure of the plasma decrease with increasing altitude, the magnetic field fills the entire space in the corona, and dominates the dynamics of the low corona. The confinement of the plasma there creates the structures shown in Fig. 2.1.

The plasma motions in the photosphere and below inject energy into the magnetic field, which is transported by field-aligned electric currents to the overlying corona, and may be temporarily stored there (Forbes 2010). As a result of this electrodynamic coupling with the photosphere and the convection zone, the corona undergoes dynamical evolution on different scales of time and energy, ranging from coronal heating to large eruptive events, the manifestations of which are flares and CMEs. They often occur together. In the following these combined events are referred to as eruptive events. When a flare is not accompanied by a CME, it is often called a confined flare.

2.3 Solar Flares: Energy Release and Radiative Signatures of Charged Particle Acceleration

From the observational viewpoint a solar flare is defined as a temporary brightening across the electromagnetic spectrum. Radio and X-ray signatures observed during a major flare are shown in Fig. 2.2. No spatial resolution is involved. The shape of the light curve depends strongly on the nature of the emitting particle population: the relatively slow and smooth time evolution in soft X-rays (top panel) comes from the heating to $T > 10^7$ K of a coronal volume located in an active region. The emission evolves relatively slowly, on minute time scales, due to the thermal inertia of the coronal plasma. Hα emission (not shown here) comes from a cooler plasma volume in the chromosphere and displays also a smooth overall evolution. Hard X-rays (second panel from top) and microwaves, on the other hand, are mostly emitted by non-thermal electron populations, that is electrons accelerated to energies of tens of keV or sometimes several MeV. These are much higher than average energies in the pre-eruptive corona (100–200 eV) or even in the hot flare plasma (a few keV). Emission from non-thermal electrons is usually spiky, in particular during the impulsive flare phase of the flare, which is the rise phase of the soft X-ray time profile. The spikiness most probably reveals the fragmentation of the acceleration process.

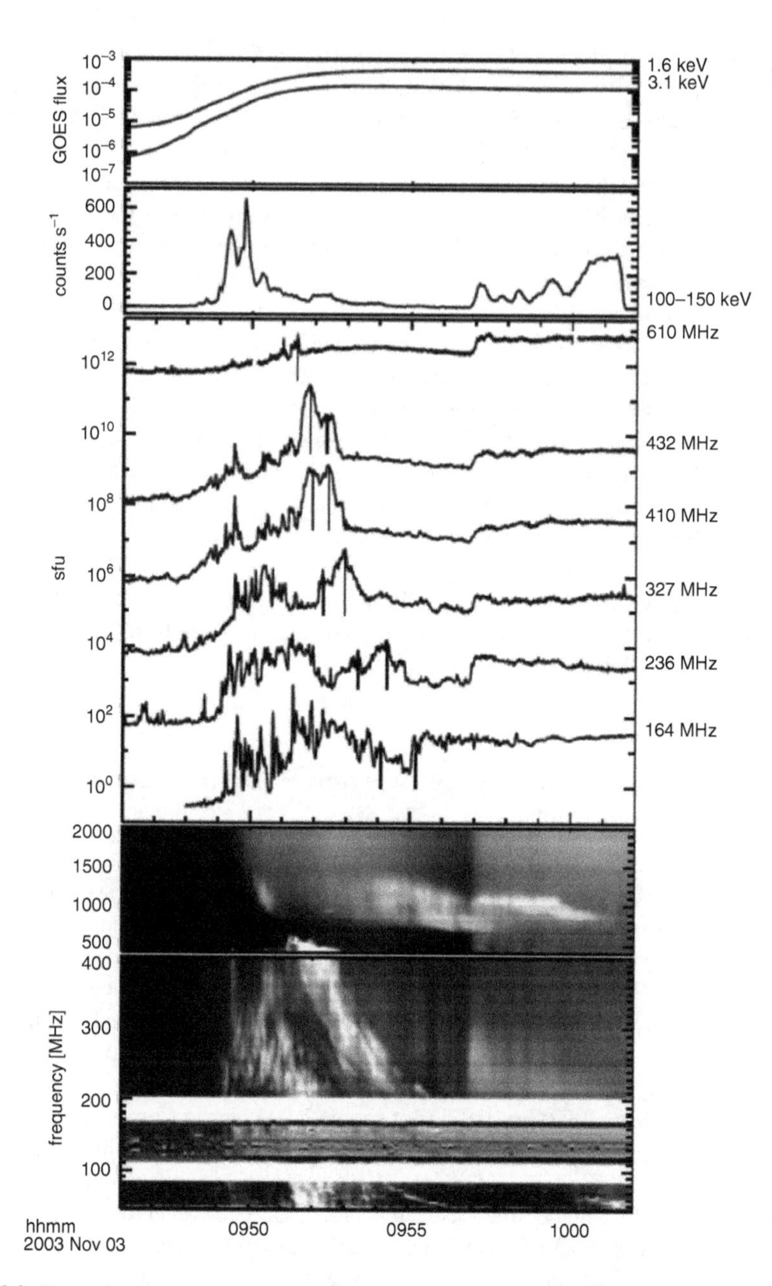

Fig. 2.2 Dynamic spectrum and single-frequency records of a complex hard X-ray and radio burst on 2003 Nov 3 (Dauphin et al. 2006). *From top to bottom*: (1) Time histories in two soft X-ray channels (GOES), (2) in one hard X-ray channel (RHESSI), (3) at six individual radio frequencies (Trieste at 610 MHz, Nançay Radioheliograph at the other frequencies), (4) dynamic spectrum from 2000 to 500 MHz (Zurich) and 400 to 40 MHz (Potsdam). Shading from *black* (background) to *white* (bright emission). *Horizontal white bands* are frequency ranges where no observation is possible because of terrestrial emitters. Credit: Dauphin et al., A&A 455, 339, 2006, reproduced with permission ©ESO

Soft X-rays are routinely observed since 1975 by the *Geostationary Operational Environmental Satellites* (GOES) operated by NOAA (USA). Because of their general availability they replaced Hα observations from ground as the standard indicator of solar flares. The reference for the importance of a solar flare is the peak soft X-ray flux in the 0.1–0.8 nm channel of GOES (also referred to as the 1–8 Angstrom channel): bursts with peak flux $n \cdot 10^{-m}$ are referred to as class X ($m = -4$), M ($m = -5$), C ($m = -6$) etc. for classes B and A, followed by a multiplier n of the order of magnitude. A flare of class X3.5 has a peak flux in the 0.1–0.8 nm band of $3.5 \cdot 10^{-4}$ W m^{-2}.

2.3.1 Emission Processes

Emission in different spectral ranges is produced by different mechanisms in different regions of the atmosphere: microwaves (1 GHz—some tens of GHz) are gyrosynchrotron radiation, while hard X-rays are bremsstrahlung of energetic electrons with ambient protons and ions. In the event shown in Fig. 2.2 a first hard X-ray burst (second panel from top, photon energies 100–150 keV) occurs during the impulsive phase of the flare. After its decay minor fluctuations persist. They are followed by a new rise near 09:57 UT. The single-frequency records at radio frequencies between 610 and 164 MHz show some similarities with the hard X-rays, especially the initial impulsive phase emission and (down to 236 MHz) the new rise near 09:57 UT. But the brightest emission at these frequencies has no obvious counterpart in the hard X-ray time profile. This emission comes from higher in the solar atmosphere than the X-rays. In fact the radio spectrum is rather complex, as shown by the dynamic spectrogram in the bottom panel. Broadband features like the late rise near 09:57 UT are clearly visible, but are a relatively faint background emission. The broadband nature points to the gyrosynchrotron mechanism as the basic process. The brightest emissions are very strongly structured in frequency and time. They reveal a variety of different radiation processes, which involve micro instabilities of the plasma that are far from being completely understood. These emissions are collectively referred to as "plasma emission".

2.3.1.1 Bremsstrahlung

Free electrons travelling through a background of ions are deflected by the Coulomb force. This change of momentum and energy is balanced by the emission of a photon. The long range of the Coulomb force implies that the electron's motion is dominated by multiple interactions at long distance within the Debye sphere. Each individual interaction creates only a small deflection, corresponding to the emission of a photon with small momentum and energy, at radio wavelengths. Close encounters are rare, but each one creates an individual strong deflection, associated with the emission of a high-energy (X-ray or gamma-ray) photon.

Since bremsstrahlung is a collisional process, the volume emissivity depends on the product of the electron number density and the ion number density, or rather on the sum of this product over protons and all ion species in the plasma. If the impacting electrons are non-thermal, with density n_e^*, the emissivity is proportional to the product with the ambient ion density, $n_e^* n_i$. In a thermal plasma, since $n_e \sim n_i$, the emissivity is $\sim n_e^2$. Thermal bremsstrahlung produces the radio emission of the quiet solar atmosphere and also a weak microwave emission in bursts. However, much higher intensities can be reached at radio frequency by other processes. Bremsstrahlung of non-thermal electrons produces the hard X-ray emission of solar flares (Tandberg-Hanssen and Emslie 1988). Because of its dependence on the ambient ion density, the emission comes from dense layers of the solar atmosphere, mostly the chromosphere.

2.3.1.2 Gyrosynchrotron Radiation

The smooth broadband emission of moderate flux density in Fig. 2.2 is ascribed to gyrosynchrotron radiation by mildly relativistic electrons (energies from about 100 keV to a few MeV). This mechanism is the generally adopted interpretation of the microwave spectrum of solar flares, implying magnetic fields of some hundreds of gauss in the low corona, and may on occasion extend to much lower frequencies.

The radiation is produced by electrons gyrating in magnetic fields. Thermal electrons in the corona, which have rather low energy, may emit at low harmonics of the electron cyclotron frequency

$$\nu_c = \frac{1}{2\pi} \frac{eB}{m_e} , \tag{2.1}$$

where e is the electric charge, m_e the mass of the electron, B the magnetic field intensity. In magnetic fields of order 1000 G $= 0.1$ T the cyclotron frequency is in the GHz range. As long as the energy of the emitting electron is low, an observer in the plane of the cyclotron motion will see an electric field that varies nearly sinusoidally in the course of time, corresponding to a frequency spectrum that shows a signal at the cyclotron frequency and its low harmonics. For a relativistic electron with speed υ the fundamental frequency of the emission is the electron cyclotron frequency divided by the Lorentz factor $\gamma = 1/\sqrt{1-(\upsilon/c)^2}$. Its radiation is strongly beamed in the direction of motion. The observer looking at a single gyrating electron will perceive a flash each time the velocity vector of the electron is along the line of sight, and will record a succession of sharp pulses. The frequency spectrum contains numerous lines at harmonics of the relativistic cyclotron frequency. In practice these lines are broadened, and the spectrum is a continuum. For a highly relativistic electron ($\gamma >> 1$; synchrotron radiation) with pitch angle α the emitted spectrum extends from the cyclotron frequency up to the critical frequency

$$\nu_c = \frac{3}{2}\gamma^2 \nu_{ce} \sin\alpha \ . \tag{2.2}$$

The emission is referred to as gyrosynchrotron in the case of mildly relativistic electrons as observed in solar flares.

2.3.1.3 Plasma Emission from Electron Beams

The radio spectrum of Fig. 2.2 shows a wealth of structure at decimetre-metre waves (~1000–40 MHz), which contrast with the smooth evolution of the gyrosynchrotron spectrum. For instance, a cluster of narrow-band features is seen between 200 and 400 MHz in the time interval 09:50–09:53. A structured band of emission drifts from about 500 MHz (09:51–09:53) down to 200 MHz (09:53–09:56). The clear spectral structure with narrow bandwidth as compared to the central frequency ($\Delta\nu/\nu \ll 1$) and the high flux density are typical of collective or coherent emission processes, where a perturbation of the plasma makes the entire electron population radiate at the characteristic frequencies of the plasma. In the solar corona the plasma frequency is the relevant frequency, because it is higher than the cyclotron frequency.

Most often plasma emission arises from some micro-instability due to deviations of the electron distribution function from isotropy. We outline this mechanism for a type III burst from a magnetic field-aligned electron beam: when a beam of field-aligned fast electrons is superposed on a Maxwellian, a positive slope arises over some velocity range in the distribution function measured along the magnetic field. The electrons in the beam interact with plasma waves with phase speeds in the range where the distribution function has a positive slope. These are Langmuir waves. If there is more energy in electrons slightly above the phase velocity than in electrons that are slower, waves are excited at the expense of the kinetic energy of the electron beam. Electrons are removed from the beam and transferred to lower speeds so that the beam distribution is flattened to a plateau. The instability ceases when the distribution function has no longer a positive slope. Actual measurements of such distribution functions in space are shown in Ergun et al (1998). A recent review is Sinclair Reid and Ratcliffe (2014).

The Langmuir waves cannot escape from the corona, but can transfer their energy to electromagnetic waves, for instance through coupling with other waves: low-frequency waves like ion sound waves or high-frequency waves, especially other Langmuir waves. The transfer of energy between waves can occur provided the three waves satisfy parametric conditions, which can be formulated like the conservation of energy ($h\nu$) and momentum in quantum mechanics. Energy conservation implies that the frequency of the resulting wave must be the sum of the frequencies of the interacting waves. The frequency of Langmuir waves ν_L is close to the electron plasma frequency. The frequency ν_{EM} of an electromagnetic wave generated through the coupling process with an ion-sound wave (frequency ν_{lf}) is $\nu_{EM} = \nu_L + \nu_{lf} \simeq \nu_L \simeq \nu_{pe}$. Coupling of two Langmuir waves yields $\nu_{EM} = \nu_L + \nu_{L'} \simeq 2\nu_{pe}$. The electromagnetic wave is hence generated near the electron plasma frequency or its harmonic. Depending on which process prevails, the resulting electromagnetic emission is referred to as "fundamental" or "harmonic". Both may occur together, or one of the two modes may prevail.

The dependence of the plasma frequency on the ambient electron density creates the typical frequency behaviour of type III bursts: as the electron beam proceeds to increasing altitude, hence to lower ambient electron density, the frequency of the emission decreases. The result is a short burst that drifts from high to low frequencies. In a hydrostatic atmosphere the frequency drift rate is directly related to the speed of the exciter along the density gradient. Relatively fast drifts in type III bursts are related to a fast exciter, namely an electron beam. The slowly drifting type II bursts are ascribed to a slower exciter, namely a shock wave. An example is the burst between 09:53 and 09:55 in the 200–400 MHz range of Fig. 2.2.

2.3.1.4 Gamma-Rays from Accelerated Protons and Ions

So far only electron-related radiative diagnostics were addressed. This is a real bias, since observational signatures of non-thermal protons and ions are more difficult to obtain in the solar atmosphere. These particles emit at gamma-ray wavelengths.

Prominent nuclear lines are produced when accelerated protons or ions with energies in the range 1–100 MeV/nucleon bombard the low solar atmosphere. The nuclei of the ambient medium are excited to high energies, and subsequently relax by emitting the gamma-ray photons. If the impacting high-energy particle is a proton and the target a heavy ion, the target does not move significantly, and the emitted gamma-ray line is narrow. If an accelerated ion hits a target proton or helium, it continues its motion and emits a Doppler-shifted line, which is considerably broadened by the angular distribution of the ions.

The most widely observed nuclear line is the neutron capture line near $h\nu = 2.2$ MeV. When protons with energies exceeding about 30 MeV interact with other nuclei, neutrons are released, initially with high energy. When thermalised after some tens of seconds, they can be captured by ambient protons to form a deuterium nucleus. Its binding energy is released via emission of a photon at 2.223 MeV. At still higher energies, above 300 MeV/nucleon, nuclear interactions of protons and helium nuclei with ambient protons create pions, which decay rapidly. Pion-decay positrons eventually annihilate with electrons. Neutral pions decay into photons and create a specific emission feature at $h\nu > 60$ MeV, which in solar gamma-ray spectra shows up as a high-energy bump on top of the decaying spectrum of electron bremsstrahlung. A recent review of solar gamma-ray emission is given by Vilmer et al. (2011). Pion decay gamma-ray emission from relativistic protons will be addressed in more detail in Chap. 8.

2.3.2 *Where Are Electrons Accelerated in Solar Flares?*

Images of flares in hard X-rays show frequently configurations that look like magnetic loops with a particle acceleration region near or above the top. In the RHESSI image of the 2005 Jan 20 flare (Fig. 2.3a) the red contours outline a thermal X-ray source at temperatures above 10^7 K, which traces the upper part of a coronal loop. The blue contours are the sources of hard X-ray emission from non-thermal electrons precipitated into the chromospheric footpoints. The two elongated grey bands onto which the footpoints project are flare ribbons seen in UV. They outline regions of the chromosphere heated by energy deposition during the flare. This source morphology is generally interpreted as a signature of energy release near or above the loop top, which heats the plasma in the coronal loop and accelerates electrons. They escape from the primary acceleration site as magnetic field-aligned beams. Because of their high energy they interact very little with the low-density coronal plasma and precipitate into the dense chromosphere at the base of the loop. They lose their energy instantaneously through collisions with this dense environment, while simultaneously emitting a small amount as hard X-rays. The RHESSI image is a snapshot: during the impulsive phase of the event the hard X-ray sources occur in an irregular temporal succession at neighbouring places on the UV ribbons. A standard cartoon scenario as in Fig. 2.3b locates the energy release above the loop top, probably related to magnetic reconnection. The upward field lines may be part of a plasmoid that is ejected upward or they may be open to the high corona. The energy release may equally well be related to magnetic reconnection

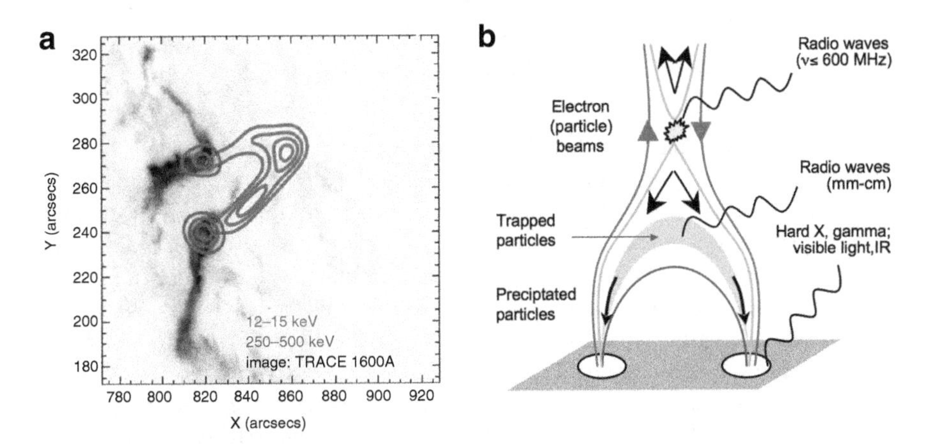

Fig. 2.3 (**a**) Contours of hard X-ray emission in two spectral bands (RHESSI) superposed on a negative TRACE image of chromospheric flare ribbons in UV (Krucker et al. 2008). Credit: Krucker et al., Astrophys. J. 678, L63 (2008). ©AAS. Reproduced with permission. (**b**) Cartoon scenario of particle populations and related electromagnetic emissions during a flare. From Klein and Dalla (2017)

with another closed magnetic structure. A more detailed discussion of hard X-ray source morphology and its interpretation can be found in Holman et al. (2011).

Radio emission of electron beams, such as type III bursts, is another key observation to identify the electron acceleration. In some flares Aschwanden and coworkers were able to identify radio emissions from bidirectional electron beams, with downward-directed beams at high frequency (high ambient electron density) and upward directed beams at low frequency. The authors concluded that these beams came from a common acceleration region where the electron plasma frequency had the value corresponding to the frequency from where the oppositely drifting radio bursts emanate. They derived an ambient density of about $(1–10) \cdot 10^9$ cm^{-3}. From the timing of peaks at different hard X-ray energies they concluded that the acceleration region is placed at a typical altitude of about 1.5 times the half-length of the magnetic loop above the photosphere (see Sects. 3.3 and 3.6 of Aschwanden 2002). This corresponds to cartoon scenarios such as Fig. 2.3b.

It is not clear if protons and ions are accelerated in the same regions as electrons. A close connection between relativistic electrons (above 300 keV) and protons above 30 MeV is suggested by the observed correlation of the peak fluxes of their respective bremsstrahlung and nuclear line emissions (Shih et al. 2009). But differences become visible in the detailed time evolution (Kiener et al. 2006), and the source locations are in general not identical (see review in Vilmer et al. 2011). While models exist to explain different acceleration regions of these particle species, for instance in terms of resonances with different types of waves, the observations provide a number of challenges which show that our understanding is far from complete.

2.3.3 A Qualitative View of Acceleration Processes

An electric field is needed to accelerate charged particles. This is simply because only the electric component of the Lorentz force $\mathbf{F} = q(\mathbf{E}+\mathbf{v}\times\mathbf{B})$ is able to change the energy, $\frac{dW}{dt} = \mathbf{v}\cdot\mathbf{F} = q\mathbf{v}\cdot\mathbf{E}$. Since the solar corona is a highly conducting medium in most parts (comparable to copper), no static electric field can be maintained along the ambient magnetic field. Peculiar configurations where transient magnetic-field-aligned electric fields can exist are current sheets and shock waves. We cannot measure electric fields in the corona, only infer them from the plasma motions.

In the cartoon scenario of Fig. 2.3b, where a reconnection region is depicted by oppositely directed magnetic fields, the most elementary electric field is the motional $\mathbf{E} = -\mathbf{V} \times \mathbf{B}$ induced by the inflow of plasma into the reconnecting current sheet. A test particle exposed to this field will $\mathbf{E} \times \mathbf{B}$-drift into the current sheet. In the region where the magnetic field is near zero, the particle decouples from the magnetic field. Protons propagate along the induced electric field, while electrons propagate opposite to it. Both particle species are hence accelerated. The stationary situation shows the principle, but is unlikely to be encountered in the solar corona. Current sheets are expected to fragment into magnetic islands. They correspond to parallel electric currents that attract each other, so that the magnetic islands formed during the fragmentation coalesce. Charged particles are trapped in a highly dynamical medium between coalescing magnetic islands and gain more energy (Cargill et al. 2012).

Reconnection jets are another ingredient of magnetic reconnection that can lead to particle acceleration. They evacuate the plasma from the reconnection region. The jets may generate shock waves when impinging on the underlying magnetic structures, or waves (turbulence) in the ambient plasma. Different particle species interact with different types of waves. This may explain preferential acceleration of some particle species, as observed in those SEP events where the ^{3}He abundance is enhanced by several orders of magnitude with respect to the quiet solar corona. Acceleration processes at shock waves and in turbulent plasmas are discussed in more detail in Chap. 3.

2.4 What Is a Coronal Mass Ejection?

The observational definition of a coronal mass ejection (CME) is an extended outward travelling feature in white-light coronographic images. This means that the visible manifestation is the outward motion of plasma. Phenomenologically many CMEs have a three-part structure in coronographic images as shown in Fig. 2.4: an outer bright region, which is understood to be mainly composed of plasma swept up from the ambient corona by the outward propagating piston, a dark cavity, which is low-density material in the ejected magnetic structure, and a bright core consisting of filament material. This basic structure of the CME is created by the magnetic field. A CME is hence the ejection of a large-scale coronal magnetic structure together with the confined plasma.

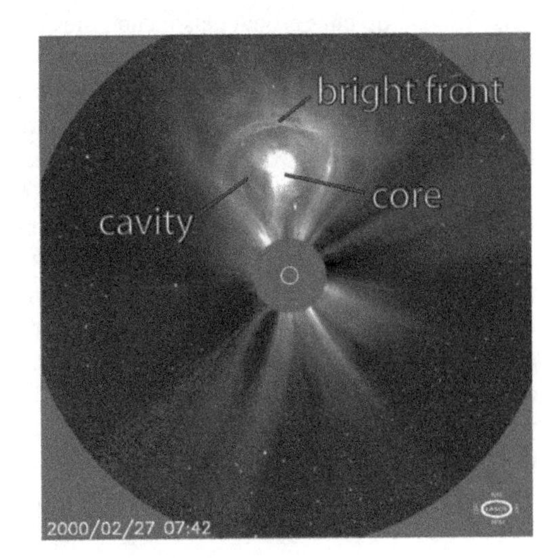

Fig. 2.4 Coronographic image of a CME in white light, from Riley et al. (2008). Credit: Riley et al., Astrophys. J. 672, 1221 (2008). ©AAS. Reproduced with permission

The time-height diagram constructed from tracking the front of the white-light feature through the field of view of the coronograph shows CMEs that travel at constant speed in the field of view, while others accelerate or decelerate. The fastest part is in general the outward-moving apex of the CME, but in some events fast lateral expansion is also observed. Outward speeds observed by SoHO/LASCO since 1996 range from a poorly defined lower limit of some tens of km s^{-1} to a maximum of 3400 km s^{-1} (Gopalswamy 2009). The CME leaves the Sun and travels through the heliosphere. The physical process behind this phenomenology is a large-scale instability of coronal magnetic structures. A recent review on CMEs is Chen (2011).

2.4.1 CME Magnetic Structure and Eruption

The magnetic structure outlined by the dark CME cavity (Fig. 2.4) looks like a closed two-dimensional magnetic field. More detailed studies suggest it is the projection of a three-dimensional magnetic flux rope, where the magnetic field lines are helices wound around the confined plasma. Such a flux rope is sketched in Fig. 2.5a. The blue-green loop-like structure is the plasma in the flux rope, the blue and black field lines indicate the helicoidal magnetic field in and around the flux rope. The Lorentz force on this configuration is directed upward, since the magnetic field lines are more densely packed below the flux rope than above. The upward Lorentz force, sometimes called "hoop force", is balanced in equilibrium by the downward-directed Lorentz force exerted by the surrounding coronal magnetic field, whose field lines are plotted in orange. An excess upward force can be generated for instance by the torsion of one foot of the flux rope and its magnetic field, due to the plasma motions in the photosphere. When this happens, the excess magnetic pressure below the flux rope is enhanced—the flux rope is lifted by the Lorentz force (Fig. 2.5b), ambient coronal plasma and the embedded magnetic field are convected from both sides towards the region where it was located before, and oppositely directed magnetic fields can reconnect. This is illustrated in Fig. 2.5b, c for two field lines, with the reconnection happening in a limited region schematically indicated by the yellow symbol of an explosion. New magnetic field is then added to the flux rope (the upper part of the field line drawn in red colour), and new magnetic loops form in the low corona. These reconnected loops appear as arcades in EUV images. Their formation may continue over several hours.

A 2D projection of this situation is depicted in Fig. 2.5d, together with the consequences of the magnetic reconnection: charged particles accelerated in transient electric fields around the reconnection region, and electromagnetic emissions excited directly or indirectly by these particles in different regions of the erupting configuration. Hard X-rays and gamma-rays are generated respectively by electrons and ions in the dense low atmosphere. Radio emission is generated by energetic electrons in different regions, including the dilute plasma in higher atmospheric layers. Typical radio signatures of electrons accelerated during flux

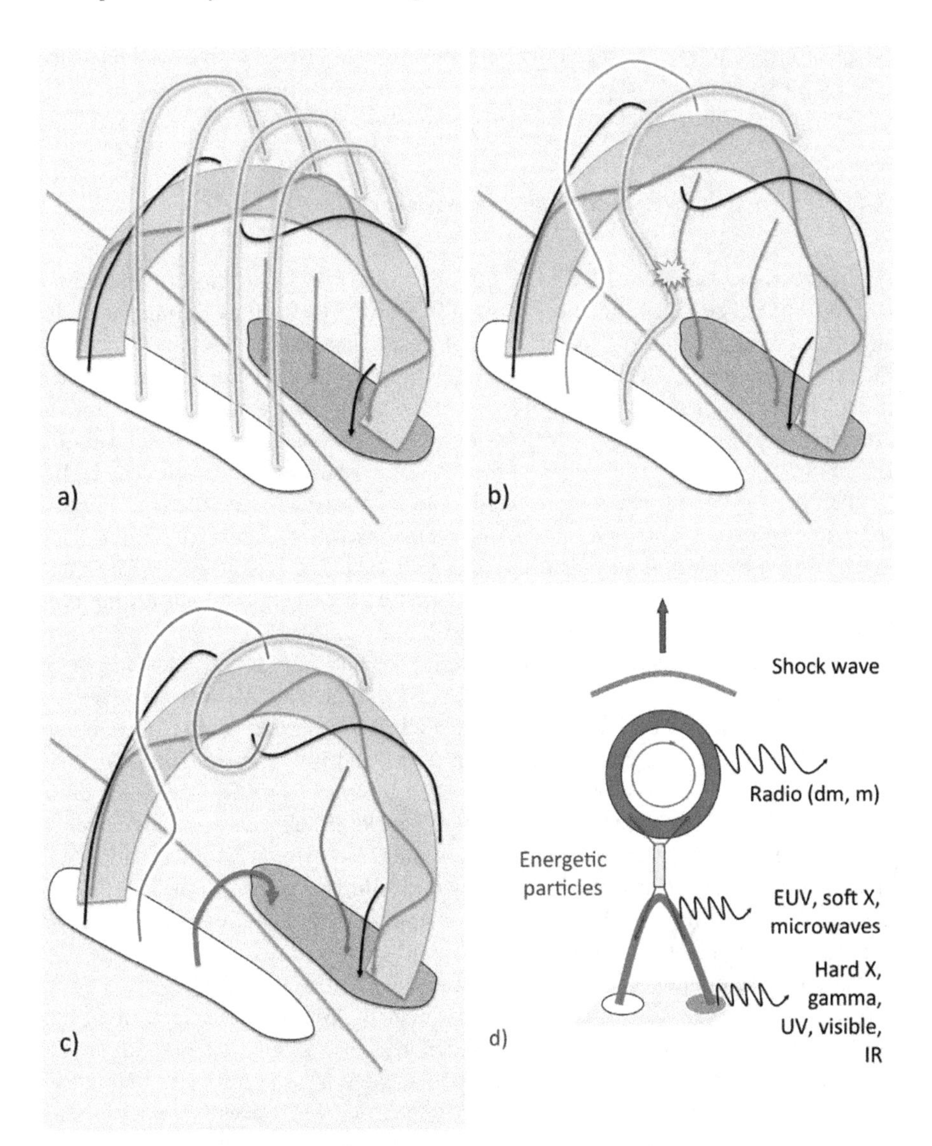

Fig. 2.5 Cartoon scenario of the magnetic field configuration around a magnetic flux rope in the solar corona (**a**), and of its evolution during the liftoff of a coronal mass ejection (CME; **b**, **c**). The *white and grey-shaded areas* indicate opposite magnetic polarities in the photosphere, separated by the *grey line*, where the vertical photospheric magnetic field is zero. Figure (**d**) shows a two-dimensional projection of (**c**)

rope eruptions are broadband continua like the one created by gyrosynchrotron emission in the bottom panel of Fig. 2.2, or similar features created by plasma emission from trapped electrons. Since the electrons are simultaneously present in a range of ambient electron densities, plasma emission occurs at a corresponding range of frequencies, i.e. over a broad spectral band. These emissions are called type IV bursts.

2.4.2 Shock Waves and Particle Acceleration at CMEs

A characteristic signature of the corona observed in EUV in the aftermath of a CME liftoff is the arcade of loops, which is thought to form as the magnetically stressed corona reconnects. The progress of the reconnection process is depicted in the cartoons of Fig. 2.5, where a loop formed in the course of reconnection is plotted in red. Both reconnection and turbulence in the aftermath of the CME are able to accelerate particles in a similar way as during the impulsive phase of solar flares. Because of the outward movement of the CME, the relevant processes occur higher up than during the impulsive flare phase, and less particles get access to the low solar atmosphere. All processes involving collisions with the ambient plasma, such as electron bremsstrahlung or nuclear line radiation, are less efficient. The main evidence on particle acceleration in the aftermath of a CME is a long-lasting type IV burst at frequencies below about 1 GHz.

Fast CMEs are likely to exceed the Alfvén speed and the fast magnetosonic speed of the coronal plasma, and to drive shock waves. Such shock waves are observed in situ in the heliosphere, and can be inferred in the corona from type II bursts at radio frequencies, for instance between 09:51 and 09:55 in the 500–100 MHz range in Fig. 2.2. Shock waves have also been inferred from UV spectroscopy in the corona (Mancuso 2011). The associated shocks usually have moderate Mach numbers: model-dependent interpretations of the radio spectra of type II bursts give $M_A = 1.2$–2.9 (Vršnak et al. 2002). But modelling of stereoscopic white-light observations (Rouillard et al. 2016) shows that higher Mach numbers can be found in localized regions of some CMEs.

Particle acceleration at coronal shock waves is discussed in Chap. 3. Two broad categories of processes are commonly distinguished: shock drift acceleration occurs predominantly in quasi-perpendicular shock regions, where the normal on the shock front and the upstream magnetic field include a large angle. In the shock-drift acceleration process particles gain energy from the convective electric field $\mathbf{E} = -\mathbf{V} \times \mathbf{B}$ in the shock frame, where $\mathbf{V}$ is the inflow speed of the plasma, $\mathbf{B}$ the upstream magnetic field in the rest frame of the shock. In the upstream region, electrons and ions $\mathbf{E} \times \mathbf{B}$-drift towards the shock front. Because the magnetic field is compressed by the oblique shock, the particles undergo a gradient drift along the shock front. This drift is directed along the electric field for positively charged particles, and opposite to the electric field for negatively charged particles. Hence electrons, protons and ions gain energy. Depending on the energy and pitch angle before the first encounter with the shock, the energy gained from the drift along the convection electric field may be such that the particle is again injected into the upstream medium and may escape. Besides a gradient drift, a particle also undergoes a curvature drift while its guiding centre travels along the magnetic field, which is curved in the shock transition. In a planar fast shock, the angle between a field line and the shock normal is larger downstream than upstream, and the curvature drift is opposite to the gradient drift. The curvature drift hence leads to energy losses. But the drift speed decreases as the shock becomes more oblique, so that in quasi-perpendicular shocks the gradient drift, and hence the energy gain, dominates.

In work on shock-acceleration of SEPs the acceleration process most often invoked is diffusive shock acceleration. When ions are reflected at shock waves and stream into the upstream region, they acquire a beam-like distribution and are therefore likely to generate waves, parallel-propagating Alfvén waves as well as obliquely propagating fast magnetosonic waves. When these waves grow to sufficient amplitudes, they can scatter subsequent ions back to the shock. Since the shock propagates faster than these waves, ions find themselves confined between approaching scattering centers downstream and upstream, and gain energy by bouncing back and forth through the shock front, until they eventually escape. In order to interact with the waves, the particles must be able to escape into the upstream medium after the initial reflection. This means that they must stream away from the shock front at a minimum speed $V/\cos\theta_{\mathrm{Bn}}$, where V is the speed of the shock as above, and θ_{Bn} the angle between the shock normal and the upstream magnetic field vector. Since this speed is the smaller, the smaller θ_{Bn}, diffusive shock acceleration is expected to work best at quasi-parallel shocks.

2.5 Summary and Conclusion

Charged particles may be accelerated during flares and CMEs to energies largely above the mean energy in the corona. The energy transferred to the particles is drawn from plasma flows in the photosphere and below, transported to the corona along magnetic field lines, and stored in the coronal field. The release involves magnetic reconnection or the loss of equilibrium of large-scale magnetic structures, leading to CMEs.

Particle acceleration processes related to magnetic reconnection, turbulence and shock waves are all supported by observations of electromagnetic emissions and SEPs. The evolution of the coronal magnetic field is usually described by magnetohydrodynamics, and the accelerated particles as test particles. High fluxes of non-thermal particles may, however, develop sufficient pressure and energy to back-react on the magnetic field configuration, invalidating the MHD and test particle hypotheses. This is especially expected in large solar events, where the energetic particles contain a substantial amount of the energy released during the flare or the CME (Emslie et al. 2005; Mewaldt et al. 2005). We are therefore far from a complete understanding of how particle acceleration proceeds.

In the present author's opinion it is not possible to identify an acceleration process that is *a priori* more plausible than others. In astrophysical settings like supernovae, shock waves produced in the course of a gravitational instability are widely considered as a privileged process to explain energetic particle populations.

Shock waves in the solar corona are always the result of plasma processes that are themselves conducive to particle acceleration. The preference for shock waves as particle accelerators is not justified under these conditions. Constraints from observations remain essential to understand how particle acceleration is related to the different manifestations of eruptive solar activity, and which physical processes are at work.

Acknowledgements The author is grateful to the HESPERIA consortium for many interesting discussions, and to J. Kiener for his comments on the manuscript.

References

Aschwanden, M.J.: Space Sci. Rev. **101**, 1 (2002). doi:10.1023/A:1019712124366

Cargill, P.J., Vlahos, L., Baumann, G., Drake, J.F., Nordlund, Å.: Space Sci. Rev. **173**, 223 (2012). doi:10.1007/s11214-012-9888-y

Chen, P.F.: Living Rev. Sol. Phys. **8**, 1 (2011). doi:10.12942/lrsp-2011-1

Dauphin, C., Vilmer, N., Krucker, S.: Astron. Astrophys. **455**, 339 (2006). doi:10.1051/0004-6361:20054535

Emslie, A.G., Dennis, B.R., Holman, G.D., Hudson, H.S.: J. Geophys. Res. (Space Phys.) **110**, A11103 (2005). doi:10.1029/2005JA011305

Ergun, R.E., Larson, D., Lin, R.P., et al.: Astrophys. J. **503**, 435 (1998). doi:10.1086/305954

Forbes, T.: In: Schrijver, C.J., Siscoe, G.L. (eds.) Heliophysics: Space Storms and Radiation: Causes and Effects, p. 159. Cambridge University Press, Cambridge (2010)

Gopalswamy, N.: In: Tsuda, T., Fujii, R., Shibata, K., Geller, M.A. (eds.) Climate and Weather of the Sun-Earth System (CAWSES) Selected Papers from the 2007 Kyoto Symposium, Tokyo: TERRAPUB, pp. 77–120 (2009)

Holman, G.D., Aschwanden, M.J., Aurass, H., et al.: Space Sci. Rev. **159**, 107 (2011). doi:10.1007/s11214-010-9680-9

Kiener, J., Gros, M., Tatischeff, V., Weidenspointner, G.: Astron. Astrophys. **445**, 725 (2006). doi:10.1051/0004-6361:20053665

Klein, K.L., Dalla, S.: Space Sci. Rev. (2017). doi:10.1007/s11214-017-0382-4

Krucker, S., Hurford, G.J., MacKinnon, A.L., Shih, A.Y., Lin, R.P.: Astrophys. J. Lett. **678**, L63 (2008). doi:10.1086/588381

Mancuso, S.: Sol. Phys. **273**, 511 (2011). doi:10.1007/s11207-011-9770-1

Mewaldt, R.A., Cohen, C.M.S., Labrador, A.W., et al.: J. Geophys. Res. (Space Phys.) **110**, A09S18 (2005). doi:10.1029/2005JA011038

Riley, P., Lionello, R., Mikić, Z., Linker, J.: Astrophys. J. **672**, 1221–1227 (2008). doi:10.1086/523893

Rouillard, A.P., Plotnikov, I., Pinto, R.F., et al.: Astrophys. J. **833**, 45 (2016). doi:10.3847/1538-4357/833/1/45

Shih, A.Y., Lin, R.P., Smith, D.M.: Astrophys. J. Lett. **698**, L152 (2009). doi:10.1088/0004-637X/698/2/L152

Sinclair Reid, H.A., Ratcliffe, H.: Res. Astron. Astrophys. **14**, 773 (2014). doi:10.1088/1674-4527/14/7/003

Tandberg-Hanssen, E., Emslie, A.G.: The Physics of Solar Flares. Cambridge University Press, Cambridge/New York (1988)
Vilmer, N., MacKinnon, A.L., Hurford, G.J.: Space Sci. Rev. **159**, 167 (2011). doi:10.1007/s11214-010-9728-x
Vršnak, B., Magdalenić, J., Aurass, H., Mann, G.: Astron. Astrophys. **396**, 673 (2002). doi:10.1051/0004-6361:20021413

Chapter 3
Particle Acceleration Mechanisms

Rami Vainio and Alexandr Afanasiev

Abstract This chapter provides a short tutorial review on particle acceleration in dynamic electromagnetic fields under scenarios relevant to the problem of particle acceleration in the solar corona and solar wind during solar eruptions. It concentrates on fundamental aspects of the acceleration process and refrains from presenting detailed modeling of the specific conditions in solar eruptive plasmas. All particle acceleration mechanisms (in the solar corona) are related to electric fields that can persist in the highly conductive plasma: either electrostatic (or potential) or inductive related to temporally variable magnetic fields through Faraday's law. Mechanisms involving both kinds of fields are included in the tutorial.

3.1 Introduction

Solar energetic particles (SEPs) are accelerated in flares and coronal mass ejections (CMEs) (see Chap. 2 and references therein). In this chapter, we will give a tutorial review of particle acceleration mechanisms in solar eruptions. The aim is not to give a comprehensive review of the original literature but rather give the reader an idea of what are the main ideas that could explain the acceleration of ions and electrons to the relativistic energies, we observe to be produced in the SEP events. For recent reviews on SEPs we direct the reader to Desai and Giacalone (2016) and Reames (2017).

SEPs are accelerated in the electromagnetic fields related to the dynamics of the solar corona and solar wind during solar eruptions. In quasi-static fields of the quiet-time solar corona, strong electric fields would not exist and particle acceleration beyond quasi-thermal energies (up to $\sim$keV) would seldom occur. However, solar eruptions are manifestations of very non-thermal conditions favorable for particle acceleration.

Magnetic reconnection is by definition a process, which requires electric fields to be present in the system. As discussed in Chap. 2 the plasma advected towards

R. Vainio (✉) • A. Afanasiev
Department of Physics and Astronomy, University of Turku, Turku, Finland
e-mail: rami.vainio@utu.fi; alexandr.afanasiev@utu.fi

O.E. Malandraki, N.B. Crosby (eds.), *Solar Particle Radiation Storms Forecasting and Analysis, The HESPERIA HORIZON 2020 Project and Beyond*, Astrophysics and Space Science Library 444, DOI 10.1007/978-3-319-60051-2_3

a current sheet with a frozen-in magnetic field will be associated with a convective electric field, which can accelerate particles. The situation is most likely a dynamic one, where induced electric fields with closed field lines will be generated by temporally varying magnetic fields. As discussed in Sect. 3.2.1, this can lead to very efficient conditions for particle acceleration. Reconnection outflows from the corona towards the Sun may generate collapsing loop systems, which can also act as particle accelerators. In addition to direct acceleration by large scale fields, reconnection jets in flares can also be turbulent regions. These can give rise to stochastic acceleration in the plasma (see Sect. 3.2.5). Even shock acceleration (Sect. 3.2.3) could in principle occur in reconnection regions, since the super-Alfvénic reconnection jets can be terminated by dense plasma or strong magnetic fields impeding their flow. Flares are commonly thought of as being the source of (at least) the impulsive SEP events (Reames 2017). These events show peculiar abundance ratios of ions difficult to explain in terms of non-selective acceleration processes, as shock acceleration. Presently the abundances are probably best explained by stochastic acceleration (Petrosian 2012).

Especially the gradual SEP events are undoubtedly related to CMEs. Mostly, but not exclusively, because of the morphology of the time-intensity profiles of the MeV and deka-MeV protons, showing clear organization based on the magnetic connectivity of the observer to the CME, the bow shock driven by the fast magnetized eruption has been identified as the prime candidate for accelerating the ions in SEP gradual events (Desai and Giacalone 2016; Reames 2017). There is a large amount of evidence that CME-driven shocks accelerate protons up to some tens of MeVs, but the case becomes more open at energies approaching and exceeding 100 MeV. Theoretical modeling (see, e.g., Chap. 9), however, suggests that CME-driven shocks are able to accelerate ions also to relativistic energies. For electrons, strong correlation between the observed flux properties of deka-MeV protons and MeV electrons (see, e.g., Chap. 7) could point towards shock acceleration as well, but the theoretical foundations of electron acceleration by shocks are not on such solid basis as for ions, mainly because electrons resonate with plasma fluctuations at much lower scales than protons (see Sect. 3.2.2). The main shock acceleration mechanisms are, however, able to accelerate electrons as well. In addition to shock acceleration, the CME downstream region hosts regions of strong plasma compression and turbulence. Such regions may accelerate particles via the compressional acceleration mechanism (Sect. 3.2.4) and stochastically (Sect. 3.2.5), respectively. Note also, that coronal magnetic restructuring (outside the flaring active region) behind the CME may also lead to particle acceleration via many of the same mechanisms.

3.2 Acceleration Mechanisms

Particle acceleration in solar plasmas requires electric fields and the ability of particles to propagate along them, since the energy gain rate produced by the

Lorentz force, $q(\mathbf{E} + \mathbf{v} \times \mathbf{B})$, is $\dot{W} = q\mathbf{v} \cdot \mathbf{E}$. Here q, $\mathbf{v}$, and W are the charge, the velocity and the kinetic energy of the particle and $\mathbf{E}$ and $\mathbf{B}$ are the electric and the magnetic field. Electric fields can be derived from the scalar and vector potential of the electromagnetic field:

$$\mathbf{E} = -\nabla\phi - \frac{\partial \mathbf{A}}{\partial t}, \tag{3.1}$$

and the magnetic field from the vector potential $\mathbf{B} = \nabla \times \mathbf{A}$. Particles can move in the direction of the electric field in several possible ways. In the following, we will discuss the different possibilities briefly.

3.2.1 Large-Scale Electric Field Acceleration

We will first discuss electric fields in large scales, meaning that the field changes over scales that are large compared to the gyroradii of the accelerated particles. In the simplest case, the electric field has a component parallel to the magnetic field, along which ions and electrons can move without being affected by the magnetic field. Large-scale, static parallel electric fields are, however, not easy to set up in dilute coronal plasmas, since in the near absence of collisions, electrons can usually move very quickly along the magnetic field and build up charge separation to counter any accelerating parallel electric field. In the presence of magnetic field inhomogeneities, however, gradient and curvature drifts can move ions and electrons in the direction perpendicular to the field. This may lead to particles gaining energy, if there is an electric field along the direction of the drift motion. This is the basis of several shock-acceleration mechanisms to be discussed in Sect. 3.2.3.

Magnetic reconnection is perhaps the most obvious plasma process to set up large-scale electric fields. Some models consider static electric fields accelerating particles near magnetic nulls, where particles can propagate along the electric field unimpeded (Litvinenko 1996). This, however, is not the only way to set up a large scale electric field that particles can utilize effectively to get accelerated.

Choosing to work in the Coulomb cauge ($\nabla \cdot \mathbf{A} = 0$), the electric fields due to charge separation are all described by the scalar potential and, thus,

$$\nabla^2\phi = -\frac{\rho_q}{\epsilon_0}. \tag{3.2}$$

Therefore, the part of the electric field related to the vector potential contains inductive fields related to the time derivative of the magnetic field via Faraday's law. These electric fields are then described by fields lines that are either closed loops or extend to the boundaries of the system. Electric fields induced by the temporally variable magnetic field (or vector potential) can sometimes lead to very efficient acceleration of particles. Let us illustrate this with an example.

Consider

$$\mathbf{B} = B_0 \left(\frac{y\mathbf{e}_x - x\mathbf{e}_y}{r_0} + \gamma_0 t \mathbf{e}_z \right) = \nabla \times \mathbf{A} \tag{3.3}$$

$$\mathbf{A} = B_0 \left(-\frac{y\mathbf{e}_x - x\mathbf{e}_y}{2} \gamma_0 t + \frac{x^2 + y^2}{2r_0} \mathbf{e}_z \right), \quad \nabla \cdot \mathbf{A} = 0 \tag{3.4}$$

$$\mu_0 \mathbf{J} = \nabla \times \mathbf{B} = -\frac{2B_0}{r_0} \mathbf{e}_z \tag{3.5}$$

This magnetic configuration is represented by a helical magnetic field, where the axial field increases linearly with time and the poloidal field (driven by a constant axial current) is constant (see Fig. 3.1).

The axial magnetic field is generated by a poloidal external current flying at some radial distance, e.g., $r = r_0$, increasing linearly with time. We neglect the displacement current so the time evolution of the axial field (and external current) has to be slow in the sense that $r_0 \gamma_0 \ll c$. With these assumptions, the induced electric field is

$$\mathbf{E} + \nabla \phi = -\frac{\partial \mathbf{A}}{\partial t} = \gamma_0 B_0 \frac{y\mathbf{i} - x\mathbf{j}}{2} = \tfrac{1}{2} r_0 \gamma_0 \mathbf{B}_\mathrm{p}. \tag{3.6}$$

Clearly, the induced electric field is in the direction of the poloidal magnetic field and, therefore, particles propagating along the field lines will be accelerated along the field, ions in one direction and electrons in the other. The axial field, however, will make sure that electrons and ions will also propagate along the axial direction,

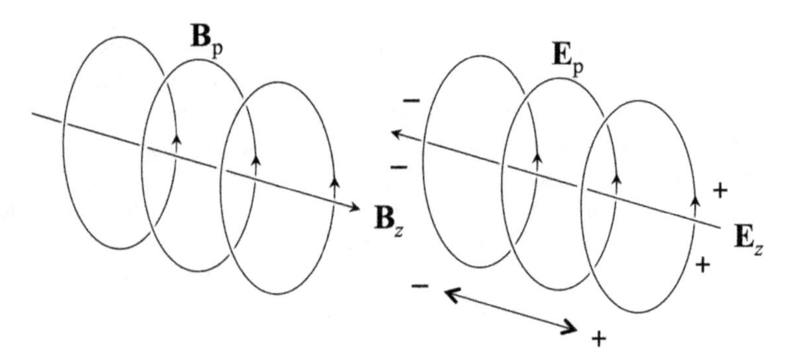

Fig. 3.1 Electromagnetic fields in the simple model. The magnetic field (*left*) is a helix, where the axial field ($\mathbf{B}_z$) is linearly increasing with time and the poloidal field ($\mathbf{B}_\mathrm{p}$) is driven by a homogeneous axial current. The poloidal electric field $\mathbf{E}_\mathrm{p}$ (*right*) is induced by the time-dependent axial magnetic field and it is in the same direction as the poloidal magnetic field. Charge separation caused by the particles accelerated by $\mathbf{E}_\mathrm{p}$ along the magnetic field lines will produce an oppositely directed axial electric field. This would eventually turn the electric field perpendicular to the magnetic field if charge separation is let to build up in the plasma

in the opposite directions. Thus, charge separation will be set up. It produces a potential field in the axial and radial direction, which combined to the induced field could produce a total electric field perpendicular to the magnetic field. The condition would be

$$\mathbf{E} \cdot \mathbf{B} = \frac{r^2 \gamma_0}{2 r_0} B_0^2 - B_0 \gamma_0 t \frac{\partial \phi}{\partial z} \sim 0$$

$$\Rightarrow \phi \sim \frac{r^2 z B_0}{2 r_0 t} \qquad \Rightarrow -\nabla \phi \sim -\frac{r z B_0}{r_0 t} \mathbf{e}_r - \frac{r^2 B_0}{2 r_0 t} \mathbf{e}_z \tag{3.7}$$

However, it is clear that this field can be approached only after a finite time (as it diverges at $t \to 0$). Thus, during a limited amount of time, the induced electric field can accelerate particles very efficiently.

We notice also that particles could be accelerated by the induced electric field even without the poloidal magnetic field in the system. If only the axial magnetic field is present, particles will perform gyro-motion around it. The contour integral of the Lorentz force along one circular Larmor orbit,

$$\Delta W_\perp = \oint_{\mathrm{L}} q \mathbf{E} \cdot \mathrm{d}\mathbf{l}, \tag{3.8}$$

is in the left-handed [right-handed] sense around the magnetic field $\mathbf{B} = B(t)\, \mathbf{e}_z$ for $q > 0$ [$q < 0$]. Taking the integral in the right-handed sense (denoted below by the minus sign after L) gives

$$\begin{aligned}\Delta W_\perp &= -|q| \oint_{\mathrm{L}-} \mathbf{E} \cdot \mathrm{d}\mathbf{l} = -|q| \int_{A_\mathrm{L}} (\nabla \times \mathbf{E}) \cdot (\mathbf{e}_z\, \mathrm{d}S) \\ &= |q| \int_{A_\mathrm{L}} \frac{\partial \mathbf{B}}{\partial t} \cdot (\mathbf{e}_z\, \mathrm{d}S) = |q| \pi r_\mathrm{L}^2 \frac{\partial B}{\partial t}\end{aligned} \tag{3.9}$$

Since the Larmor radius is $r_\mathrm{L} = v_\perp \tau_\mathrm{L} / 2\pi$, where $\tau_\mathrm{L} = 2\pi \gamma m / |q| B$ is the particle gyrotime, and γ and m its Lorentz factor and mass, respectively, we have

$$\begin{aligned}v_\perp \Delta p_\perp &= \Delta W_\perp = |q| \pi \frac{v_\perp \tau_\mathrm{L}}{2\pi} \frac{\gamma m v_\perp}{|q| B} \frac{\partial B}{\partial t} \\ \Rightarrow \frac{\dot{p}_\perp}{p_\perp} &= \frac{1}{p_\perp} \frac{\Delta p_\perp}{\tau_\mathrm{L}} = \frac{1}{2B} \frac{\partial B}{\partial t} = \tfrac{1}{2} \gamma_0,\end{aligned} \tag{3.10}$$

which describes the betatron acceleration process. Clearly, $p_\perp = p_{\perp 0} \mathrm{e}^{\frac{1}{2} \gamma_0 t}$ gives the perpendicular particle momentum in terms of its initial value $p_{\perp 0}$ at $t = 0$.

3.2.2 *Resonant Wave Acceleration*

As discussed above, generating electric fields with quasi-static components along the magnetic field is not easy. Instead of large scale fields, the fields can also be at the gyromotion scale, and such fields can be carried by various plasma waves. At first sight, if the electric field is fluctuating, it is quasi-periodically pointing in opposite directions and the time integral of the acceleration would seem to vanish in most cases. In this case, however, it is possible for the particle to be in resonance with the wave's electric field: if the period of particle motion agrees with the period of the wave, the phase of the wave at the location of the particle can be constant, leading to a constant accelerating electric field felt by the particle.

This process can be illustrated with a simple linearized model. Consider the electric field of a circularly polarized wave propagating along the mean magnetic field taken to be constant and along the z axis. Thus,

$$\mathbf{E}_1 = E_1[\mathbf{e}_x \cos(kz - \omega t) - \mathbf{e}_y \sin(kz - \omega t)], \tag{3.11}$$

and the phase speed of the wave is $V_\phi = \omega/k$. The electric and magnetic fields of the wave are related via Faraday's law, i.e.,

$$\begin{aligned}
\frac{\partial \mathbf{B}_1}{\partial t} &= -\nabla \times \mathbf{E}_1 = -\begin{vmatrix} \mathbf{e}_x & \mathbf{e}_y & \mathbf{e}_z \\ \partial_x & \partial_y & \partial_z \\ E_{1x} & E_{1y} & 0 \end{vmatrix} = \mathbf{e}_x \frac{\partial E_{1y}}{\partial z} - \mathbf{e}_y \frac{\partial E_{1x}}{\partial z} \\
&= kE_1[-\mathbf{e}_x \sin(kz - \omega t) - \mathbf{e}_y \cos(kz - \omega t)] && (3.12) \\
\Rightarrow \mathbf{B}_1 &= \frac{k}{\omega} E_1[-\mathbf{e}_x \cos(kz - \omega t) + \mathbf{e}_y \sin(kz - \omega t)] && (3.13) \\
\Rightarrow \mathbf{E}_1 &= -\frac{\omega}{k} \mathbf{B}_1 = -V_\phi \mathbf{B}_1, && (3.14)
\end{aligned}$$

i.e., the magnetic field is circularly polarized, as well. In case the phase speed is less than the speed of light, we can always make a boost along the z axis to a coordinate system where the wave frequency is zero. In this frame the wave electric field vanishes and the charged particles interacting with the wave conserve their energy. In the laboratory frame the change in the particle energy is, thus, obtained through

$$\Delta W = \Gamma_\phi V_\phi p' \Delta\mu', \tag{3.15}$$

where $\Gamma_\phi = (1 - V_\phi^2/c^2)^{-1/2}$, μ' and p' are the pitch-angle cosine and the (constant) momentum magnitude of the particle in the wave frame, and ΔX represents the change in the quantity X. The equations of motion in the wave frame (omitting the primes) can be written as

$$\dot{\mu} = \frac{B_1}{B_0}\sqrt{1-\mu^2}\Omega_0 \cos(\varphi + kz) \tag{3.16}$$

$$\dot{\varphi} = -\Omega_0 + \frac{B_1}{B_0}\frac{\mu}{\sqrt{1-\mu^2}}\Omega_0 \sin(\varphi + kz) \tag{3.17}$$

$$\dot{z} = v\mu, \tag{3.18}$$

where φ is the phase angle measured around the z axis and $\Omega_0 = qB_0/\gamma m$ is the particle's relativistic (signed) gyro-frequency. Now, assuming that the magnetic amplitude of the wave is small compared to the mean magnetic field, $B_1 \ll B_0$, we can approximate the phase angle evolution by $\varphi = \varphi_0 - \Omega_0 t$. Substituting that to the equation for μ shows a resonance (a constant right-hand side) at $kz - \Omega_0 t = 0$

$$\Rightarrow kv\mu = \Omega_0, \tag{3.19}$$

which is the cyclotron resonance condition between charged particles and electromagnetic waves propagating parallel to the mean magnetic field, written in the rest-frame of the wave. Transforming back to the laboratory frame gives the resonance condition as

$$\omega - kv\mu = -\Omega_0, \tag{3.20}$$

where negative (positive) frequencies denote left-handed (right-handed) polarization. The circular polarization of the wave has to match the handedness of the particle motion in the guiding-centre rest-frame (where $v\mu = 0$), so ions (electrons) resonate with left-hand (right-hand) polarized waves in that frame. The wave-particle resonance is sketched for protons and cold-plasma waves in Fig. 3.2. We can see that low-energy protons (with $|v\mu| \lesssim 2V_A$) can only resonate with counter-propagating left-handed waves (Vainio 2000).

Approximating the left-handed dispersion curve with a straight line at $-\Omega_0 < \omega < 0$ simplifies the picture so that the phase speed of the resonant wave is constant. Thus, the relevant wave-frame, where the proton would conserve its energy, is propagating at speed $\pm V_A$ along the background field and protons with positive [negative] parallel speed resonate with waves with $\omega/k_\parallel = -V_A$ $[\omega/k_\parallel = +V_A]$. Thus, their position in $(v_\parallel, v_\perp)$ velocity plane is constrained on semicircles centered at $v_\parallel = \pm V_A$. Starting from low speeds, $v_0 \ll V_A$, wave-frame pitch-angle scattering leads to an increase of perpendicular velocity of protons, as depicted in Fig. 3.3. If the extent of the initial distribution in the parallel direction is v_0, then the extent of the final distribution in the perpendicular direction would be

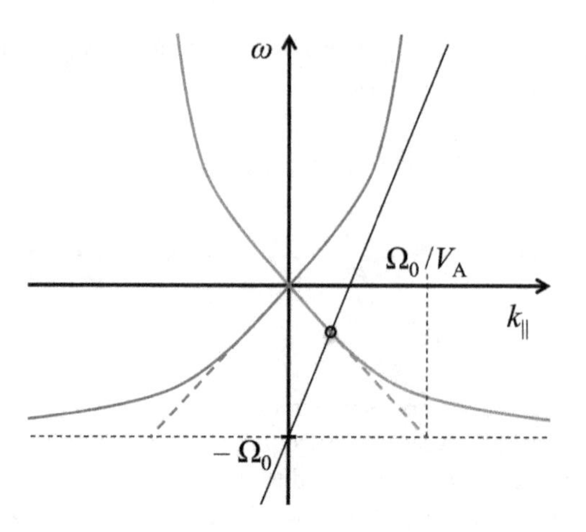

Fig. 3.2 Cyclotron resonance of protons with parallel propagating cold-plasma waves. The resonance condition, Eq. (3.19), is depicted with the *black line* for a proton with positive parallel speed $v\mu$. The *green and the red curves* give the dispersion relations of parallel-propagating cold-plasma waves with negative frequencies denoting the left-handed Alfvén–ion-cyclotron waves and positive frequencies the right-handed fast-MHD–whistler waves. The *red (green) line* gives the wave with positive (negative) phase speed $\omega/k_{\parallel}$. The *red (green) dashed lines* depict the low-frequency approximation to the dispersion relation, i.e., the Alfvén waves with positive (negative) phase speed, $\omega/k_{\parallel} = \pm V_{\rm A}$. The wave-particle resonances are found where the *black line* crosses the dispersion curves

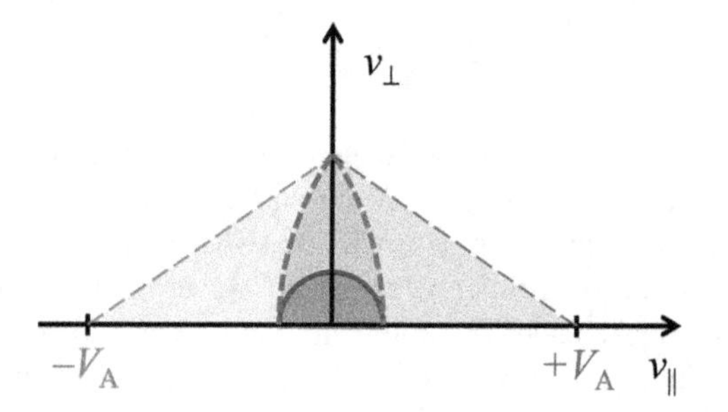

Fig. 3.3 The resonant-wave acceleration process. Particles are scattered off electromagnetic waves propagating in the medium at phase speeds $\pm V_{\phi}$, here approximated with $\pm V_{\rm A}$. Scattering is elastic in the wave frame, which leads to particles conserving their kinetic energies in that frame. If waves have left-handed polarization, low-energy ions resonate with them when propagating in the opposite direction (in the plasma frame) than the wave. Particles initially in the *grey region* inside the *blue semi-circle* will be scattered by the waves to fill the intersection of the *green and red sectors*, centered at $v_{\parallel} = -V_{\rm A}$ and $v_{\parallel} = +V_{\rm A}$, respectively

$\sqrt{(V_{\rm A}+v_0)^2 - V_{\rm A}^2} = \sqrt{2V_{\rm A}v_0 + v_0^2}$ and the ratio of final perpendicular and initial parallel energies is

$$\frac{W_{\perp,\max}}{W_{\parallel,0}} = \frac{v_0^2 + 2V_{\rm A}v_0}{v_0^2} = 1 + \frac{2V_{\rm A}}{v_0}, \tag{3.21}$$

so obviously the mechanism alone cannot accelerate ions to very high energies. This mechanism is, however, the basis of the cyclotron heating models of the solar corona (Isenberg 2001).

The resonant wave-acceleration process in solar flares is thought to be responsible for the preferential acceleration of minor ions. The dispersion relations of the waves in a multi-species plasma are not as simple as in Fig. 3.2, but contain additional resonances ($|k_\parallel| \rightarrow \infty$) with thermal He^4 ions at cyclotron frequency half of protons, $\Omega_\alpha = \frac{1}{2}\Omega_p$. Ions with cyclotron frequencies differing from that of protons and alpha particles are much more efficiently accelerated than protons and alphas in such plasmas. We will come back to this in Sect. 3.2.5.

3.2.3 Shock Acceleration

Shocks can accelerate particles in various ways. The three most commonly studied mechanisms are shock drift acceleration (SDA), shock surfing acceleration and diffusive shock acceleration (DSA).

SDA occurs, when a particle interacts once with a quasi-perpendicular shock front (Fig. 3.4, panels a and b) (Sarris and Van Allen 1974). The particle drifts (due to the motional electric field, $\mathbf{E} = -\mathbf{u}_1 \times \mathbf{B}_1$) with the upstream bulk speed u_{1x} toward the shock wave. When the ion [electron] hits the shock front, it feels the stronger downstream magnetic field, meaning that its Larmor radius is smaller than in the upstream field, and that its guiding center shifts parallel [anti-parallel] to the electric field. Thus, the particle is accelerated. During their interaction with the shock the particles, at least in an averaged sense, conserve their first adiabatic invariant, $p_\perp^2/B$. For a perpendicular shock, we then get

$$\frac{p_2^2}{B_2} = \frac{p_1^2}{B_1} \quad \Rightarrow \quad p_2 = p_1\sqrt{\frac{B_2}{B_1}}, \tag{3.22}$$

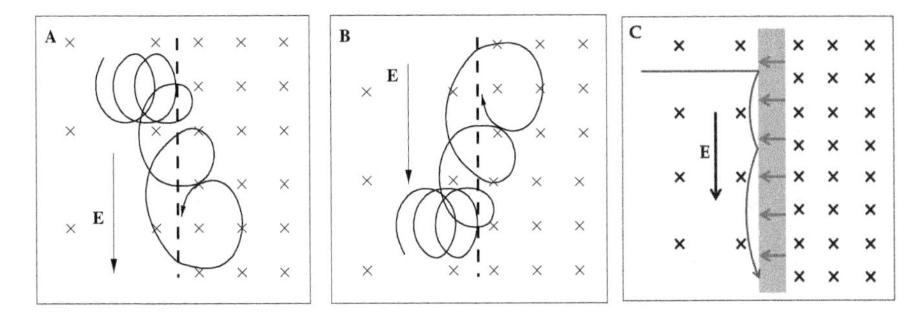

Fig. 3.4 Shock drift acceleration and shock surfing. An energetic charged particle is convected to a quasi-perpendicular shock from upstream by the electric-field drift. In the shock front, due to magnetic field gradient in the front, (**a**) ions drift parallel and (**b**) electrons drift anti-parallel to the electric field and, thus, gain energy. Panel (**c**) depicts an ion surfing on the shock due to multiple reflections by the cross-shock potential electric field. (Panels (**a**) and (**b**) from Koskinen 2011)

so the particle momentum in the downstream region, p_2, is approximately $\sqrt{B_2/B_1}$ times the particle momentum in the upstream region, p_1. The total gain in energy is, therefore, not very significant, but may explain the so-called shock spike events observed during nearly perpendicular interplanetary shocks crossing the spacecraft.

SDA operates in oblique shocks as well, but slightly modified. It is most advantageous to transform to the so-called de Hoffmann–Teller frame, where the flow and the magnetic field are parallel throughout the shock and the motional electric field vanishes. This transformation from the upstream plasma frame is along the magnetic field at speed $u_1 = u_{1x}/\cos\Theta_{\rm Bn}$ away from the shock in the upstream region. Here $\Theta_{\rm Bn}$ is the shock normal angle. Now, particles incident on the shock from the upstream side with high-enough pitch-angle, i.e., with $1-\mu^2 > B_1/B_2$, cannot enter the downstream side of the shock while conserving their adiabatic invariant and, thus, are reflected back to the upstream region. This reflection process, when viewed in the upstream plasma frame, give the reflected particle a parallel momentum addition of the same order as transmission in the quasi-perpendicular shock.

Shock surfing (Fig. 3.4, panel c) is related to SDA in the sense that the accelerating electric field is the same, the motional electric field. However, the drift motion along the field is related to the upstream-directed cross-shock potential electric field, caused by charge separation at the shock front (Shapiro and Üçer 2003). The cross-shock potential electric field can specularly reflect ions incident on the shock and enable their acceleration in the direction of the motional electric field. The specular reflections can occur multiple times at a grazing angle. This leads to ion trajectories surfing on the shock front in the direction of the electric field. The maximum energy gained in the process depends on the thickness of the shock and can be estimated to reach the MeV range at interplanetary shocks, if the shock thickness is as low as an electron skin depth (Koskinen 2011). Note that electrons will not be accelerated in this simple type of model, which has a monotonic potential (and, thus, unidirectional field) inside the shock front.

One encounter with the shock does not typically lead to a very substantial gain of energy. If, however, particles can interact with the shock many times, acceleration becomes more efficient. Shock surfing is not the only way this can happen. Particles can interact with magnetic irregularities in the plasma flow, and this can change the particle's propagation direction relative to the shock front enabling several encounters with the shock. Since particle transport under such conditions is described by diffusion relative to the local plasma flow, this acceleration mechanism is called diffusive shock acceleration (DSA) (Drury 1983; Lee 1983).

DSA can be best understood by considering shock waves propagating parallel to the magnetic field. There, as the particle crosses the shock front, its velocity vector does not change, because the magnetic field is not compressed. When the particle is moving relative to the plasma under the influence of frozen in magnetic turbulence, providing scattering centers,[1] it conserves its energy in the local plasma frame while

[1]For simplicity, the turbulence in the vicinity of the shock is often assumed to be magnetostatic, but the DSA theory can be formulated assuming that the scattering centers are propagating Alfvén waves. This simply modifies the velocities of the scattering centers.

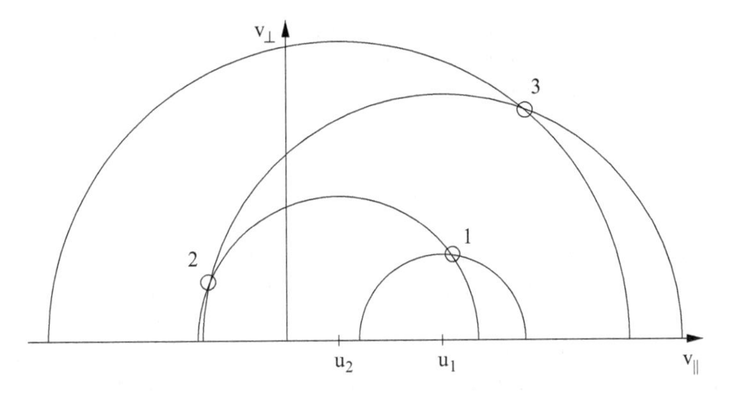

Fig. 3.5 Diffusive shock acceleration. An energetic charged particle scatters off magnetic irregularities frozen in to the local plasma flow. The numbered points depict successive crossings of the shock front, where the speed of the scattering centers changes. Because points with odd numbers must have $v_{\|} > 0$ and points with even numbers must have $v_{\|} < 0$, the shock crossings lead to a systematic gain of energy $W = \frac{1}{2}mv^2$. (From: Koskinen 2011)

simultaneously scattering in pitch angle. Upstream [downstream] particles are, thus, staying on semicircles in velocity space, centered at $(v_{\|}, v_{\perp}) = (u_{1[2]}, 0)$. Due to pitch-angle scattering, energetic ($v' > u$) particles can propagate in either direction relative to the shock. When the flow at the shock is compressed (i.e., $u_2 < u_1$), particles crossing the shock many times gain speed systematically as shown in Fig. 3.5.

When particle speeds are much larger than the fluid speeds, $v \gg u$, particle distributions become almost isotropic as a result of the scattering process. This enables one to calculate the energy spectrum of accelerated particles resulting from DSA. After n shock crossings, the mean particle momentum is

$$\langle p_n \rangle = p_0 \exp\left\{ \frac{4}{3} \sum_{j=1}^{n} \frac{u_1 - u_2}{v_j} \right\}, \tag{3.23}$$

where $p_0 \gg mu_1$ is the injection momentum. The probability of a particle of performing at least n crossings of the shock is

$$P_n = \exp\left\{ -4 \sum_{j=1}^{n} \frac{u_2}{v_j} \right\} = \left(\frac{\langle p_n \rangle}{p_0} \right)^{-3u_2/(u_1-u_2)}. \tag{3.24}$$

By combing these, the integral momentum spectrum can be given as

$$N(p > \langle p_n \rangle) = N_0 \left(\frac{\langle p_n \rangle}{p_0} \right)^{-3/(r-1)}, \tag{3.25}$$

where N_0 is the total number of particles injected to the acceleration process and $r = u_1/u_2$ is the compression ratio of the shock. Thus, shock-accelerated particles have a power-law differential momentum spectrum

$$\frac{\mathrm{d}N}{\mathrm{d}p} = \frac{3N_0}{r-1}\left(\frac{p_0}{p}\right)^{(r+2)/(r-1)} \tag{3.26}$$

with spectral index $\sigma = \mathrm{d}\ln N/\mathrm{d}\ln p = (r+2)/(r-1)$ solely determined by the compression ratio of the shock. While the calculation above was presented for parallel shocks, the final result applies for oblique shocks as well.

The spectral index is actually determined by the shock's compression ratio only if $M_\mathrm{A} \gg 1$. If the Mach number of the shock is of the order unity, the magnetic scattering centers in the flow (which are actually magnetohydrodynamic waves or turbulence and not static) are no longer static magnetic fluctuations. Instead, they have non-negligible phase speeds $V_\phi \sim V_\mathrm{A}$ relative to the flow. Recall that the scattering is elastic in the frame of the propagating magnetic structure, where the electric field of the fluctuation vanishes. Taking these considerations into account when determining the *scattering-center compression ratio* of the shock (Vainio and Schlickeiser 1999), one gets

$$\sigma = \frac{r_\mathrm{sc}+2}{r_\mathrm{sc}-1}; \quad \text{with} \quad r_\mathrm{sc} = \frac{u_{1x}+V_{\phi 1}}{u_{2x}+V_{\phi 2}} \stackrel{M\to\infty}{\longrightarrow} \frac{u_{1x}}{u_{2x}} = r. \tag{3.27}$$

In slow-mode shocks, the scattering centers (Alfvén waves) always have larger phase speeds than fluid speeds. Thus, the scattering centers do not converge in slow shocks under many circumstances. In such cases, DSA is not operating at the shock.

Upstream of fast-mode shocks, particles can generate their own scattering centers by streaming instabilities of the Alfvén waves (Lee 1983; Afanasiev et al. 2015). This can be figured out most easily by looking at particle scattering in the upstream plasma frame. Particles conserve their energies in the frame co-moving with the MHD waves, because in that frame, the wave has no electric field (since δB has no time dependence there, as discussed above). Thus, particle energy in the plasma frame is

$$W = W' \pm V_{\mathrm{A}1} p'_\parallel, \tag{3.28}$$

where the signs denote waves propagating forward ($+V_{\mathrm{A}1}$) and backward ($-V_{\mathrm{A}1}$) relative to the flow. Particles entering the upstream region from downstream have $p'_\parallel/W' < -(u_1 \pm V_{\mathrm{A}1}) < 0$. Scatterings make the particles isotropic in the wave frame, so as a result of scatterings in the upstream region, $p'_\parallel$ increases. Thus, particle energy in the plasma frame increases (decreases) in scatterings off forward (backward) MHD waves in the upstream region. Since the total energy of particles and waves has to remain constant in the plasma frame, this means that the energy density of backward waves increases and forward waves decreases.

Finally, one should note that the power-law spectrum does not extend to infinite energies, but experiences a cut-off at some high energy determined by the age and the size of the system. Obviously, if there is limited time τ available to accelerate the particles, they cannot be accelerated beyond energies determined by $\dot{p} \sim p/\tau$, where $\dot{p}$ is the momentum gain rate related to the scattering rates and flow velocities in the system. Likewise, when the particle's diffusion length, κ/u_1, becomes of the order of the system size, the particle can not be accelerated any further, as it will not be confined to the vicinity of the shock anymore but may escape from the system. Here, $\kappa = \frac{1}{3}\lambda v$ is the spatial diffusion coefficient and λ is the scattering mean free path, which cannot be smaller than the Larmor radius of the particle. In an inhomogeneous magnetic field, such as the coronal field, particles also need to be confined by turbulence near the shock strong enough to avoid the escape by adiabatic focusing (Vainio et al. 2014). This leads to yet another condition that $\kappa/L < u_1$, where $L = -B/(\partial B/\partial s)$ is the focusing length, i.e., the scale height of the magnetic field intensity. This represents the relevant system size in the coronal medium.

3.2.4 *Compressional Acceleration and Collapsing Magnetic Traps*

DSA operates because of the convergence of the flow of scattering centers at the shock. The acceleration rate of an isotropic population of particles in a converging flow is given by $\dot{p} = -\frac{1}{3}p\nabla \cdot \mathbf{u}$, where $\mathbf{u}$ is the scattering center velocity. Compressional acceleration can, therefore, work also in presence of compressions ($\nabla \cdot \mathbf{u} < 0$) of non-shock type. If the diffusion length of the particles, κ/u, is much longer than the gradient scale of the flow, $L \sim u/|\nabla \cdot \mathbf{u}|$, then the compression acts on the particles as a shock, i.e., the resulting spectrum is practically the same as in the case of DSA (Jokipii and Giacalone 2007). In the opposite case, $\kappa \ll u^2/|\nabla \cdot \mathbf{u}|$, the compression will accelerate the particle distribution adiabatically. Because their diffusion length is very small, particles are primarily advected through the compression and all particles regardless of their initial momentum will gain the same factor in momentum.

Let us illustrate this with a simple example. Assume that scattering centers are frozen-in in the plasma flow. Thus, $\nabla \cdot \mathbf{u}$ is given by the hydrodynamic conservation of mass as

$$\nabla \cdot \mathbf{u} = -\frac{1}{\rho}\frac{\mathrm{d}\rho}{\mathrm{d}t}, \tag{3.29}$$

where ρ is the density of the plasma parcel being advected at velocity $\mathbf{u}$. Thus,

$$\frac{\dot{p}}{p} = -\tfrac{1}{3}\nabla \cdot \mathbf{u} = \frac{\dot{\rho}}{3\rho} \Rightarrow \frac{p^3}{\rho} = \text{constant}, \tag{3.30}$$

which is consistent with the adiabatic equation of state for a monoatomic gas, i.e., $T\rho^{-2/3}$ = constant at non-relativistic energies.

Similar to compressional acceleration, particles can be accelerated adiabatically if they are confined to a collapsing magnetic trap (see Borissov et al. 2016 and references therein). A simple example is a shrinking magnetic bottle, consisting of two magnetic mirrors in the ends, e.g., a contracting coronal loop. Particles mirroring in the ends of the trap would get accelerated, as in the rest frame of the center of the trap (the loop apex in the simple example), the mirrors in the two ends of the trap would be approaching each other. The acceleration rate is obtained from the conservation of the second adiabatic invariant, $\oint p_\parallel \mathrm{d}s_\parallel = 2|p_\parallel|s = \text{constant}$, where the integral is along the magnetic field lines from one end of the trap to the other and back, and s is the length of the trap along the field. Decreasing s has to be compensated by increasing $|p_\parallel|$ and the parallel momentum will increase at rate

$$\frac{\mathrm{d}|p_\parallel|}{\mathrm{d}t} = -|p_\parallel|\frac{\dot{s}}{s} \tag{3.31}$$

As a magnetic bottle cannot trap particles residing in the loss-cone of the weaker magnetic mirror, this mechanism would appear at first sight to be limited to rather modest gains of momentum. However, if the magnetic field inside the trap is simultaneously increasing the betatron effect will increase the perpendicular momentum and help the particles stay trapped. The details of the magnetic configuration and its evolution will determine the acceleration efficiency of a collapsing magnetic trap.

3.2.5 *Stochastic Acceleration*

DSA, compressional acceleration, and collapsing traps are all examples of so-called first-order Fermi acceleration, where particles gain momentum systematically and proportionally to the speed of a moving magnetic structure. Stochastic acceleration, or second-order Fermi acceleration, is a process, where particles gain or lose energy (with a positive net energy gain) by interacting simultaneously with plasma disturbances with different phase speeds in the laboratory frame (Miller 1998). For example, if a high-energy proton is propagating in a medium with counter-propagating Alfvén waves, it can simultaneously scatter off waves propagating in both directions (Fig. 3.6, left panel). As the scatterings conserve the energy of the proton in each wave frame, the scatterings off one wave produce energy change along different characteristics in the $(v_\parallel, v_\perp)$ plane than the scatterings off the other mode. This leads to random walk of the particle in the $(v_\parallel, v_\perp)$ plane, which can be described by momentum diffusion. In this process the net momentum gain rate is proportional to the second power of the wave speed. (Hence, second-order Fermi acceleration.)

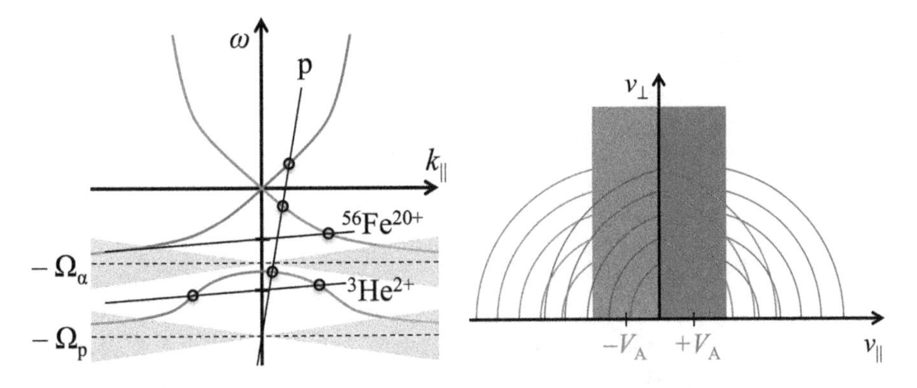

Fig. 3.6 The *left panel* shows the resonance plot for a multi-species plasma with protons and alpha particles as the major ion species. The wave modes with positive (negative) phase speeds are plotted with *green (red) curves*. The *grey shaded regions* are those where resonances with thermal ions will damp the wave power efficiently. Cyclotron waves have two branches with resonances with alpha particles and protons, respectively. The *lines* depict the resonance conditions for energetic protons ($|v\mu| \gg V_A$) and thermal minor ions. The *right panel* shows the possible velocity-space trajectories of protons in a plasma with counter propagating Alfvén waves, only. The *red (green) semicircles* are trajectories of particles interacting with Alfvén waves with positive (negative) phase speeds. The *red (green) region* depicts the velocities at which resonances with and Alfvén waves with positive (negative) phase speeds are not possible. Outside these regions, protons can always resonate with waves propagating in both directions leading to random walk in velocity space, i.e., momentum diffusion

The situation is illustrated in Fig. 3.6, right panel, where energetic protons can interact with two Alfvén waves propagating in opposite directions, when the proton velocity is situated in the region $|v_{\parallel}| \gtrsim 2V_A$. Scatterings off waves with positive (negative) phase speed will result in motion along the red (green) semi-circles showing that the particle energy in the plasma frame can change in a random fashion, i.e., increase or decrease. As this diffusive process spreads the distribution of protons in energy, the net effect will be acceleration. This is an example of stochastic acceleration, but other types of waves as well as randomly moving coherent structures interacting with the particles can lead to a similar situation of particles diffusing in momentum when interacting with the structures.

Figure 3.6, left panel also shows the resonant interaction of low-energy minor ions (^{3}He and iron) with higher-frequency left-handed wave modes. The figure shows that while thermal protons and alpha particles would typically resonate with heavily damped wave modes (and their resonant wave acceleration should, therefore, be somewhat inefficient), minor ions not only resonate with much less damped waves but also simultaneously with several waves propagating at different phase speeds. Especially ^{3}He can be very efficiently accelerated stochastically in this process, starting already from thermal energies. Stochastic acceleration process, therefore, has the power to explain the peculiar abundances (strong increase in ^{3}He and heavy ion abundances over protons and alphas) observed in impulsive events (cf. Chap. 2). Note that stochastic acceleration may also occur in the turbulent sheath regions of coronal shocks and help to explain the observed double-power-law spectral form often observed in large gradual SEP events (Afanasiev et al. 2014).

3.3 Concluding Remarks

The basic acceleration mechanisms at play in erupting coronal plasmas accelerating particles to the highest energies have been described. By making several simplifications the aim has been to convey the principles of the most important SEP acceleration mechanisms, fostering efforts to create realistic and comprehensive models of solar eruptions also from the particle acceleration aspect. The models should involve realistic descriptions of both the macroscopic and microscopic fields in the plasma, as electric fields of practically all scales from the kinetic to the global may contribute to the acceleration of particles in solar eruptions.

References

Afanasiev, A., Vainio, R., Kocharov, L.: Astrophys. J. **790**, 36 (2014). doi:10.1088/0004-637X/790/1/36

Afanasiev, A., Battarbee, M., Vainio, R.: Astron. Astrophys. **584**, 81 (2015). doi:10.1051/0004-6361/201526750

Borissov, A., Neukirch, T., Threlfall, J.: Solar Phys. **291**, 1385 (2016). doi:10.1007/s11207-016-0915-0

Desai, M., Giacalone, J.: Living Rev. Sol. Phys. **13**, 3 (2016). doi:10.1007/s41116-016-0002-5

Drury, L'O.C.: Rep. Progr. Phys. **46**, 973 (1983). doi:10.1088/0034-4885/46/8/002

Isenberg, P.A.: J. Geophys. Res. **106**, 29249 (2001). doi:10.1029/2001JA000176

Jokipii, J.R., Giacalone, J.: Astrophys. J. **660**, 336 (2007). doi:10.1086/513064

Koskinen, H.: Shocks and Shock Acceleration. In: Physics of Space Storms, pp. 279–298. Springer, Heidelberg (2011)

Litvinenko, Y.E.: Astrophys. J. **462**, 997 (1996). doi:10.1086/177213

Lee, M.A.: J. Geophys. Res. **88**, 6109 (1983). doi:10.1029/JA088iA08p06109

Miller, J.A.: Space Sci. Rev. **86**, 79 (1998)

Petrosian, V.: Space Sci. Rev. **173**, 535 (2012). doi:10.1007/s11214-012-9900-6

Reames, D.V.: Solar Energetic Particles – A Modern Primer on Understanding Sources, Acceleration and Propagation. Lecture Notes in Physics, vol. 932, 127 pp. Springer, Cham (2017)

Sarris, E.T., Van Allen, J.A.: J. Geophys. Res. **79**, 4157 (1974). doi:10.1029/JA079i028p04157

Shapiro, V.D., Üçer, D.: Planet. Space Sci. **51**, 665 (2003). doi:10.1016/S0032-0633(03)00102-8

Vainio, R.: Astrophys. J. Suppl. Ser. **131**, 519 (2000). doi:10.1086/317372
Vainio, R., Schlickeiser, R.: Astron. Astrophys. **343**, 303 (1999)
Vainio, R., Pönni, A., Battarbee, M., et al.: J. Space Weather Space Clim. **4**, A08 (2014). doi:10.1051/swsc/2014005

Chapter 4
Charged Particle Transport in the Interplanetary Medium

Angels Aran, Neus Agueda, Alexandr Afanasiev, and Blai Sanahuja

Abstract The scenario and fundamentals of the physics of charged particle interplanetary transport are briefly introduced. Relevant characteristics of solar energetic particle (SEP) events and of the interplanetary magnetic field are described. Next, the motion of a charged particle and the main assumptions leading to the description of the focused and diffusive particle transport equations utilised in the next chapters are discussed. Finally, two different models are applied to interpret SEP events.

4.1 Introduction

4.1.1 Energetic Particles in the Solar System

Major solar eruptive phenomena, solar flares and coronal mass ejections (CMEs), are usually accompanied by outbursts of charged particles that have been accelerated up to several hundred MeV/nucleon, in some instances up to a few GeV/nucleon. These solar energetic particle (SEP) events are mostly composed of ionised H with $\sim$10% He and $<$1% heavier elements. The acceleration, injection, and propagation of SEPs from their source to an observer in interplanetary space have been investigated over the last decades by a combination of in situ (space) and remote-sensing observations. SEP acceleration processes are described in Chap. 3. Here the focus is on the transport processes of SEPs in interplanetary space.

SEP events are usually classified into two types: impulsive events and gradual events. Impulsive events last for hours, are rich in electrons, ^{3}He and heavy ions, have relatively high charge states, and are produced by solar flares. Gradual events can last for days, are electron poor, have relatively low charge states, and are

A. Aran (✉) • N. Agueda • B. Sanahuja
Dep. de Física Quàntica i Astrofísica, Institut de Ciències del Cosmos (ICCUB), Universitat de Barcelona, c. Martí i Franquès, 1, E-08028 Barcelona, Spain
e-mail: angels.aran@fqa.ub.edu; agueda@fqa.ub.edu; blai.sanahuja@ub.edu

A. Afanasiev
Department of Physics and Astronomy, University of Turku, Turku FI-20014, Finland
e-mail: alexandr.afanasiev@utu.fi

O.E. Malandraki, N.B. Crosby (eds.), *Solar Particle Radiation Storms Forecasting and Analysis, The HESPERIA HORIZON 2020 Project and Beyond*, Astrophysics and Space Science Library 444, DOI 10.1007/978-3-319-60051-2_4

associated with coronal and interplanetary shocks driven by CMEs that move rapidly through the solar wind plasma. Gradual SEP events are more intense (i.e., much higher intensities and higher fluences) than impulsive events. Hybrid SEP events have also been observed; they show mixed characteristics which partially correspond to impulsive and gradual events, suggesting that solar particle events can have distinct components (i.e., flare-accelerated particles and shock-accelerated particles).

Impulsive events are generally limited to within a 30° longitude band around the footpoint of the nominal field line magnetically connected to their parent active region. On the other hand, gradual events are able to produce much wider longitudinal distributions due to the extended nature of the propagating interplanetary shock. In this way, observers located at different places in the heliosphere can be magnetically connected to different parts of the front of such a shock by interplanetary magnetic field (IMF) lines. In gradual events, in addition to the various transport effects at play, the shape of the particle intensity temporal profiles depends also on both the dynamic evolution of the shock strength and the relative location of the observer with respect to the traveling shock. The point along the shock front at which successive magnetic field lines connect with the observer is termed the cobpoint (connecting with the observer point, see Heras et al. 1995); and the gradual SEP event intensity-time profiles recorded in interplanetary space can be interpreted in terms of the cobpoint evolution, after deconvolving the transport effects. The SEP propagation in interplanetary space is controlled by the large-scale structure of the magnetic field and the turbulent magnetic field fluctuations superposed on the mean magnetic field.

4.1.2 The Interplanetary Magnetic Field

At a certain distance from the Sun, the solar wind flow speed is much higher than the local sound and Alfvén speeds. Near the Earth's orbit (1 AU), typical values for the sound and aflvénic speeds are $\sim$60 km s^{-1} and $\sim$40 km s^{-1}, respectively, whereas for the solar wind speed is $\sim$400 km s^{-1}. This implies that the plasma dynamic pressure is much higher than both the magnetic and thermal pressures (see more details in e.g., Hundhausen 1995). The solar wind carries the magnetic field from the Sun to interplanetary space, with the magnetic field *frozen-in* in the nearly radially expanding solar wind flow, given the very high conductivity of the solar wind plasma (for a deduction of the *frozen-in* condition see e.g., Bittencourt 2004). Field lines can then be seen as stream lines of the fluid flow. However, the situation becomes complicated because of the solar rotation which has an average period of 27.3 days. One can interpret the radially outflowing plasma streams as parcels of plasma emitted from the same source region. These parcels carry the magnetic field with them, and because they are tethered to the rotating Sun, the IMF lines that trail behind them are spirals. By assuming a constant solar wind speed, u, an IMF line depicts an Archimedean spiral (known as the Parker spiral after the model proposed by Parker 1958).

The equation of the Archimedean spiral can be derived from the displacement in radial and angular directions. Assuming as initial conditions of a plasma parcel on the Sun, a source longitude ϕ_0, and a source radius r_0, at time t, the parcel is found in the equatorial plane at position, r

$$r = r_0 + u(\phi - \phi_0)/\Omega \tag{4.1}$$

where Ω is the solar rotation speed. The angle between the radial direction and the magnetic field **B**, ψ, is given by $\tan\psi = r/a$ where $a = u/\Omega$. Figure 4.1 shows a sketch of an IMF line. Assuming a solar wind speed of 400 km s^{-1}, $\psi \sim 45°$ at 1.0 AU. The spiral configuration represents a smooth average of the large-scale IMF. Variations in solar wind velocity and processes acting in the solar wind and corona, including reconnection, create a spread in directions around the spiral angle. In fact, individual vectors can point to any angle superposed on this average field because of small-scale random fluctuations and, in addition, individual field lines can meander relative to the average direction.

The path length, $z(r)$, along the spiral can be estimated from $\mathrm{d}z = \sec\psi \mathrm{d}r$:

$$z(r) = \frac{a}{2}\left[\frac{r}{a}\sqrt{1+\frac{r^2}{a^2}} + \ln\left(\frac{r}{a} + \sqrt{1+\frac{r^2}{a^2}}\right)\right] \tag{4.2}$$

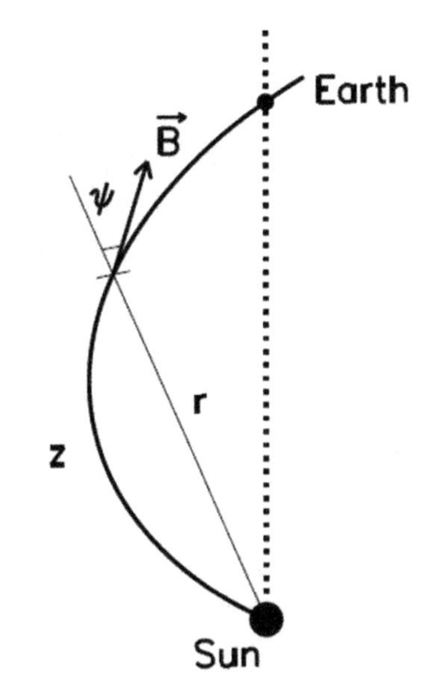

Fig. 4.1 Sketch of an Archimedean spiral interplanetary magnetic field line crossing the Earth. ψ is the angle between the radial direction and the magnetic field **B**

close to the Sun (i.e., for $r/a \ll 1$), $z \sim r$, and well beyond 1 AU (i.e., in the limit for $r/a \gg 1$), $z \sim r^2/2a$.

By assuming that the *frozen-in* condition holds in interplanetary space and that IMF lines are Archimedean spirals, the strength of the IMF is given by:

$$B(r) = B_0(r_0/r)^2\sqrt{(1+(r/a)^2)} \tag{4.3}$$

where r_0 is the heliocentric radial distance at which the field is completely frozen into the solar wind, being $r_0 > 2\,R_\odot$ and $B_0 = B(r_0)$. Thus, close to the Sun, $B(r)$ decreases as r^{-2} while well beyond 1 AU it decreases as r^{-1}. Figure 4.2 shows the dependence of the magnetic field line length and the magnetic field strength with the radial distance from the Sun.

Depending on the magnetic polarity of the photospheric footpoint of the field lines, the magnetic field spirals outward (positive) or inward (negative) from the Sun. The global interplanetary positive-negative magnetic domains are separated by a huge electric current system, the heliospheric current sheet (HCS). The HCS is tilted out of the solar equatorial plane a few tens of degrees. Changes in the coronal magnetic field due to solar activity are carried outward to space, and manifest as spatial and temporal variations (e.g., magnetic sectors). This overall picture of the IMF is known as the *ballerina skirt* model. As the HCS rotates along with the Sun, the peaks and troughs of the skirt pass through the Earth's magnetosphere, interacting with it. A detailed description of the global heliospheric magnetic field can be found in Smith (2008).

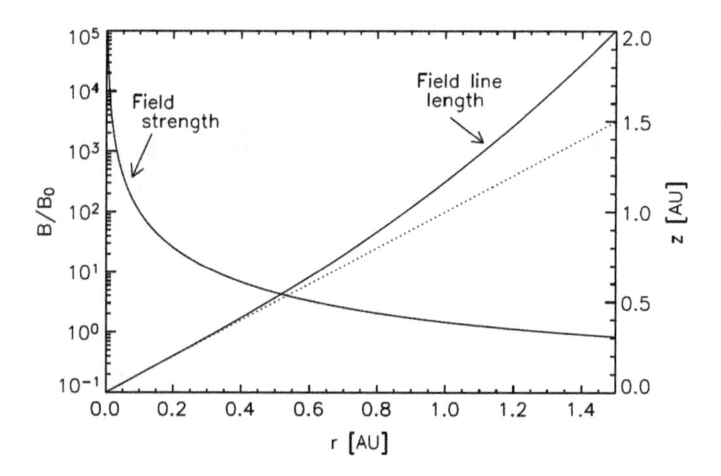

Fig. 4.2 Dependence of the magnetic field line length and the magnetic field strength with the radial distance from the Sun. The *dotted line* indicates the length of a radial line for comparison

4.1.3 Motion of Charged Particles. First Adiabatic Invariant

Space and time variable magnetic fields, with a large variety of characteristic lengths and times, play a key role in the description of the transport of SEPs in interplanetary space. The solar wind is a collisionless plasma highly conductive in which SEPs propagate basically tracking the IMF. It is also assumed here that SEPs are not able to modify the external, non-uniform and time-varying, magnetic fields. This section shortly reviews the most relevant aspects of electrodynamics that will later be used in Sect. 4.2 when introducing the transport of SEPs in interplanetary space. An extended description can be found elsewhere (e.g., Bittencourt 2004; Kallenrode 2004).

Let's consider a charged particle of rest mass m, charge q, moving with velocity $\mathbf{v}$, in a given electric field, $\mathbf{E}$, and in a magnetic field, $\mathbf{B}$. Neglecting collisions and any other external force (e.g., a gravitation force), the motion of the particle is governed by the Lorentz force

$$\frac{\mathrm{d}\mathbf{p}}{\mathrm{d}t} = q(\mathbf{E} + \mathbf{v} \times \mathbf{B}) \tag{4.4}$$

where $\mathbf{v} = \mathbf{p}/(\gamma m)$ and $\gamma = (1 + p^2/(mc^2))^{1/2}$ is the Lorentz factor of the particle.

Then, the Larmor (or gyration) radius of the particle is given by

$$r_L = \frac{v_\perp}{\omega_c} = \frac{p_\perp}{|q|B} \tag{4.5}$$

Where $v_\perp$ and $p_\perp$ are, respectively, the components of the velocity and momentum perpendicular to the magnetic field. The gyro frequency of the motion is $\omega_c = |q|B/(\gamma m)$.

To give some numbers, at 1 AU, for a typical IMF strength value ($B = 5$ nT), the Larmor radius of a proton with a kinetic energy of 700 MeV is $r_L \sim 9 \times 10^5$ km ($\sim$2.3 times the mean distance between the Earth and the Moon).

Each particle describes an helicoidal motion that can be decoupled into motions parallel ($\mathbf{v}_\parallel$) and perpendicular ($\mathbf{v}_\perp$) to the magnetic field. The parallel component describes the motion of the centre of gyration of the particle along $\mathbf{B}$ while $\mathbf{v}_\perp$ consists of a gyration motion (characterised by ω_c) around $\mathbf{B}$ plus a drift velocity component, $\mathbf{v}_F$, also perpendicular to $\mathbf{B}$. Then the particle motion is described by the gyration and the motion of the guiding centre, which is the addition of the motion parallel to $\mathbf{B}$ and a drift.

A charged particle moving under the presence of time-varying or non-uniform fields is affected by a number of velocity drifts: plasma drift (e.g., associated with an electric field), the gravitational drift, field line curvature and gradient drifts (associated with the magnetic field), or the polarisation drift (when there is a time-varying electric field). The drift velocity term, $\mathbf{v}_F$, represents the drift velocities caused by the aforementioned fields (see detailed descriptions of the various particle drifts in e.g., Bittencourt 2004; Kallenrode 2004).

The pitch angle, α, gives the relative size of the perpendicular and parallel components of the particle velocity, it is given by

$$\alpha = \tan^{-1}(v_{\perp}/v_{\parallel}) \tag{4.6}$$

In the description of SEP transport, the cosine of the pitch angle, $\mu = \cos\alpha$, is frequently used.

The magnetic moment, **m**, of a charged particle moving in a magnetic field is a measure of the magnetic flux traversing the circular section defined by the particle's Larmor radius. The kinetic energy can be written as the sum of its parallel and perpendicular components to the magnetic field, $W = W_{\parallel} + W_{\perp}$. It can be shown (e.g., Bittencourt 2004) that

$$|\mathbf{m}| = \frac{v_{\perp}p_{\perp}}{2B} = \frac{W_{\perp}}{B} = \frac{W\sin^2\alpha}{B} \tag{4.7}$$

Note that in a static uniform magnetic field, $\mathbf{v}_{\parallel}$ is constant, so the particle moves at constant velocity along **B**, and W and $W_{\parallel}$ are constant. Hence, it follows that, $W_{\perp}$ and $v_{\perp}$ are also constants of the motion.

The first adiabatic invariant can be derived from the equation of motion. It states that the magnetic moment of a particle, $|\mathbf{m}|$, is constant when moving in a slowly varying magnetic field (there are also other conditions for specific scenarios; e.g., no wave-particle interactions or that ω_c does not go through zero).

A particle moving into a converging magnetic field increases its $W_{\perp}$ while its $W_{\parallel}$ decreases, to keep $|\mathbf{m}|$ and W constant; hence, its Larmor radius decreases and its pitch angle increases. The opposite is true if the magnetic field strength decreases (e.g., diverging IMF).

In the absence of parallel electric fields (W constant) the pitch angle of a particle at two locations, with respective magnetic field strength B_1 and B_2, must satisfy:

$$\frac{\sin^2\alpha_2}{\sin^2\alpha_1} = \frac{B_2}{B_1} \tag{4.8}$$

If the magnetic field becomes strong enough, then $\mathbf{v}_{\parallel} = 0$, the direction of the particle is reversed. The particle's speed increases in the direction of decreasing **B** by the parallel component of the gradient force of the field, thus, the particle is reflected from the region of converging field lines. The parallel component of the average Lorentz force is called the mirror force because it leads to mirroring trajectories for a particle in a converging field. This effect is relevant for SEP propagation in interplanetary space. For example, it can explain that, for certain SEP events, solar- or near Sun- accelerated particles are later observed travelling back to the Sun (e.g., Tan et al. 2009).

If a particle is in a region of space between two high magnetic field regions, then the particle may be reflected at one side, travel towards the second, and also reflect there. Thus the particle motion is confined to a certain region of space, bouncing back and forth between two regions of large field strength. Examples of such a magnetic trapping scenario are the bidirectional SEP events sometimes observed in the downstream region of interplanetary shocks.

The same force that causes mirroring in a converging magnetic field, causes particle focusing in a diverging (decreasing) magnetic field. In this latter case, the particle will describe orbits with increasingly larger r_L (i.e., the pitch angle tends to approach zero). This focusing process is particularly important in the context of SEP transport given that energetic particles travel from Sun to Earth along the divergent interplanetary magnetic field. If no particle-diffusion process is considered, an isotropic energetic particle population released at the Sun will appear to come in a very narrow cone pitch angle ($\sim 1°$ wide) when observed at 1 AU.

4.2 Particle Transport

The solar wind is a collisionless plasma, hence SEPs mainly experience the effect of the electromagnetic fields. In the interplanetary space, one can assume that the average electric field vanishes due to the large conductivity of the solar wind plasma. The interplanetary magnetic field is turbulent (waves and fluctuations may be treated as perturbations added to the large-scale magnetic field configuration) and as such, it scatters particles in pitch-angle. Given the different scales involved, the transport of particles is often described by the evolution of the particle distribution, i.e., in an ensemble-averaged manner. The main objective of this section is to provide the key elements to contextualise the interplanetary models used later in this book.

4.2.1 Particle Transport Equations

The particle distribution function, f, is defined so that the number of particles $\mathrm{d}N$ in a phase space volume $(\mathbf{r} + \mathrm{d}^3\mathbf{r}, \mathbf{p} + \mathrm{d}^3\mathbf{p})$ at a time t is given by

$$\mathrm{d}N = f(\mathbf{r}, \mathbf{p}, t)\mathrm{d}^3\mathbf{r}\mathrm{d}^3\mathbf{p} \tag{4.9}$$

The phase-space volume element as well as the number of particles are Lorentz invariants, and therefore, the phase space density, $f(\mathbf{r}, \mathbf{p}, t)$, is an invariant. The Boltzmann equation is the fundamental equation of motion in the phase space. For a particle of charge q and mass m under an external electromagnetic force, it reads

$$\frac{\partial f}{\partial t} + \mathbf{v} \cdot \nabla f + q(\mathbf{E} + \mathbf{v} \times \mathbf{B}) \cdot \frac{\partial f}{\partial \mathbf{p}} = \left(\frac{\partial f}{\partial t}\right)_s \tag{4.10}$$

where the first term of the right hand side does not represent collisions but may describe the scattering of particles by electromagnetic waves and random magnetic field fluctuations. Given the random nature of the scattering processes, Eq. (4.10) is a Fokker-Planck equation. A thorough derivation of the transport equations in different scenarios can be found in e.g., Zank (2014). Another approach to study SEP transport is to start from the Vlasov equation that describes self-consistently the non-linear coupling between particles and fluctuating wave fields (e.g., Kallenrode 2004). It is widely used in plasma physics and has the same form as Eq. (4.10) but without the right hand side term. A discussion on the relation of the Vlasov and the Boltzmann equations can be found in e.g., Bittencourt (2004).

4.2.2 Focused Transport

For describing the transport of SEPs in the interplanetary space, the frame of the focused transport is the most adequate. In the focused transport model, energetic particles are considered to undergo pitch-angle scattering due to small scale irregularities in the IMF, and focusing and mirroring due to the large-scale weakening of the IMF at increasing distance from the Sun. Focused transport for SEP particles is mainly a competition between the focusing effect and the pitch-angle scattering processes. The standard equation derived by Roelof (1969) from Eq. (4.10) is:

$$\frac{\partial f}{\partial t} + v\mu\frac{\partial f}{\partial z} + \frac{1-\mu^2}{2L}v\frac{\partial f}{\partial \mu} = \frac{\partial}{\partial \mu}\left(D_{\mu\mu}\frac{\partial f}{\partial \mu}\right) \tag{4.11}$$

where f is a function of the spatial coordinate, z, measured along the IMF line, the particle momentum, p, the pitch-angle cosine, μ, and time t. v is the velocity of the particles, $D_{\mu\mu}$ is the pitch-angle diffusion coefficient describing the stochastic forces, and L is the focusing length which involves spatial variations of the guiding field (assumed to be static), given by $L(z) = -B(z)/(\partial B/\partial z)$. The second term of the left hand side of Eq. (4.11) describes the streaming of the particles along the IMF; the third term, the focusing of particles and the term in the right hand side, accounts for the scattering in pitch-angle.

In order to describe SEP events, a particle source term, Q, is added in the right hand side of Eq. (4.11) to account for the injection of particles by either a fixed source (like flares) or a mobile source (i.e., coronal/interplanetary shock waves).

By following the quasi-linear theory (QLT), the interaction between particles and waves is treated here to the first order only. The irregularities of the IMF are considered to be sufficiently small so that several gyrations of a particle are needed to modify significantly its pitch-angle. Also, it is assumed that the magnetic field irregularities can be described as waves of axially-symmetric transverse components, with wave vectors parallel to the average field (known as the 'slab model'). The combination of QLT with the slab model is known as the 'standard model' of particle scattering. The magnetic field fluctuations can be described by a power-density spectrum, $P(k) \propto k^{-\bar{q}}$ where $\bar{q}$ is the spectral slope. Under these assumptions $D_{\mu\mu}$ can be expressed analytically as Jokipii (1971):

$$D_{\mu\mu} = \frac{\nu_0}{2}|\mu|^{\bar{q}-1}(1-\mu^2), \tag{4.12}$$

where $\nu_0 = 6v/[2\lambda_{\parallel}(4-\bar{q})(2-\bar{q})]$ and provides the relation between the diffusion coefficient and the particle's mean free path parallel to the IMF, $\lambda_{\parallel}$. Values of $\bar{q}$ have been determined from observations in the range $1.3 \leq \bar{q} \leq 1.9$ with an average value of $\bar{q} = 1.63$ (Kunow et al. 1991). More details of the standard model and other equivalent descriptions for $D_{\mu\mu}$ can be found in e.g., Agueda and Vainio (2013).

Once the form of $D_{\mu\mu}$ is fixed, the other main parameter of the transport models is $\lambda_{\parallel}$. For protons, there is a dependence of the mean free path on the magnetic rigidity of the particles, $R = pc/q$, where c is the speed of light. For the standard model, $\lambda_{\parallel} \propto R^{2-\bar{q}}$, as long as $\bar{q} < 2$ (Hasselmann and Wibberenz 1970).

As is generally done, in the models by Pomoell et al. (2015) used in Chap. 9 and by Agueda et al. (2008) used in Chap. 10, the IMF is modelled as an Archimedean spiral. In this case, the radial mean free path of the particles, λ_r is related to $\lambda_{\parallel}$ by $\lambda_r = \lambda_{\parallel}\cos^2\psi$.

In the description above, transport of particles perpendicular to the average magnetic field is neglected, because in the inner heliosphere (1) particles move much smaller distances per time unit in the perpendicular direction than in the parallel direction e.g., Bieber et al. (1995) and (2) in the case of a mobile particle source, the continuous injection (during days) of shock-accelerated particles has a stronger contribution in shaping the SEP intensity profiles than cross-field diffusion does.

4.2.3 Diffusive Transport

When the spatial scales are larger than the particle's mean free path and the scattering time is small compared to the time scales of the phenomena under study, the standard diffusion equation can describe the variation of the particle distribution function. In this assumption, the spatial diffusion tensor, κ, reflects the effects of the fluctuations of the magnetic field (Jokipii 1971). In 1965, Parker (1965) was the first to describe the evolution in space of the cosmic-ray distribution using the following Fokker-Planck equation:

$$\frac{\partial f}{\partial t} + \mathbf{u}\cdot\nabla f - \frac{1}{3}p(\nabla\cdot\mathbf{u})\frac{\partial f}{\partial p} = \nabla\cdot(\kappa\cdot\nabla f)\ , \tag{4.13}$$

This equation describes the effects of spatial diffusion, advection due to the movement of the scattering centres with the solar wind, and energy changes of the particles. Equation (4.13) is only valid when the pitch-angle distribution of the particles is nearly isotropic. This is the assumption made in Chap. 9 to describe the transport of (quasi)relativistic protons from shocks low in the corona towards the Sun. However, Parker's equation is not applicable when the anisotropy of the distribution of particles is large (Jokipii 1987), as generally happens in the case of SEP events associated with interplanetary shocks e.g., Heras et al. (1995).

4.3 Application: Description of Solar Energetic Particle Events

The main application of interplanetary transport models is for the description of SEP events. First, different methods to solve the transport equations and approaches to perform data fitting are presented. Next, two interplanetary transport models, the Shock-and-Particle (SaP) model by Pomoell et al. (2015), used in Chap. 9, and the inversion model by Agueda et al. (2008), used in Chap. 10, are applied to briefly describe the observations needed for the modelling of SEP events and the derivation of the main transport parameters.

4.3.1 Numerical Techniques

Different numerical techniques can be applied to solve transport equations. Particularly, the interplanetary transport models used in Chap. 9 (i.e., Lario et al. 1998) and in Chap. 10 (i.e., Agueda et al. 2008) employ finite-difference and Monte Carlo methods, respectively. Both have advantages and drawbacks, and therefore there are models in the literature that utilise either methods or even a combination of the two. The main advantage of finite-difference methods (e.g., Ruffolo 1995; Lario et al. 1998; Dröge 2000) is that they are computationally fast, whereas the main advantage of Monte Carlo methods (e.g., Agueda et al. 2008; Afanasiev and Vainio 2013) is that tracking of individual particles is allowed by them.

Also, numerical models may follow two methods for data fitting: forward and inversion modelling. Forward models, like SaP (Pomoell et al. 2015), are inductive and based on the prediction of measurements with a given set of model parameters. On the other hand, inversion models, like SEPinversion (Agueda et al. 2008, 2012), are deductive and make use of the measurements to infer the actual values of the model parameters, and hence, no a priori assumption about the particle injection profile, Q, is needed.

4.3.2 Observations

SEP events are intensity enhancements above a background level detected typically for an extended particle energy range. The particle intensities obtained for a particular energy window are called differential intensities, J, or $\mathrm{d}J/\mathrm{d}E$. Differential intensities, with units of $[(\text{MeV sr s cm}^2)^{-1}]$, are related to the particle distribution function by $J = p^2 f$, where p is the particle's momentum.

The top panel of Fig. 4.3 shows a gradual SEP event starting on 2000 April 4, with 0.115–101 MeV proton differential intensity enhancements measured by the ACE and SOHO spacecraft. The source of this particle event was a travelling CME-driven shock that originated near to the solar west limb (e.g., Pomoell et al. 2015). This panel exemplifies the energy dependence of the intensity-time profiles observed during a single SEP event. Whereas at high energies (>30 MeV) proton intensities show a rapid onset, a maximum peak intensity, followed by a slow decay with

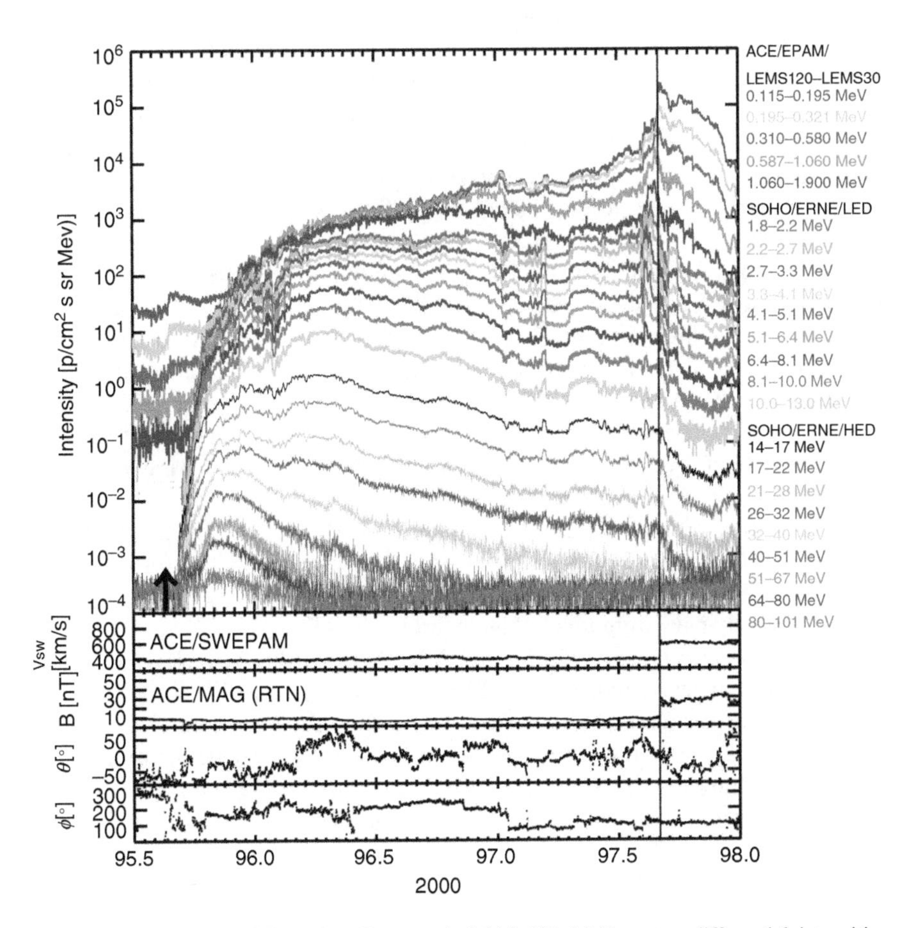

Fig. 4.3 April 4, 2000 SEP event. *Top panel*: 0.115–101 MeV proton differential intensities measured by 23 energy channels (*colour coded*) of different detectors (ACE/EPAM (Gold et al. 1998) and SOHO/ERNE (Torsti et al. 1995)). The *second panel* shows the solar wind speed measured by ACE/SWEPAM (McComas 1998) and the three *bottom panels* show the magnetic field strength, latitude and longitude (in RTN coordinates) recorded by ACE/MAG (Smith et al. 1998). The *solid vertical line* marks the time of the shock passage by the ACE spacecraft and the *arrow marks* the onset time of the associated CME

intensities being already at background level prior to the shock passage (the vertical solid line in Fig. 4.3), low energy (<2 MeV) intensities keep increasing with a marked peak at the shock. The smooth transition of the shape of flux profiles from high to low proton energies suggests that the efficiency of the shock at accelerating particles gradually diminishes with energy as it propagates away from the Sun.

Information of the solar wind and IMF is needed in order to perform the modelling and to know how the assumptions made in the models comply with the actual conditions. The bottom panels of Fig. 4.3 show smooth profiles for the solar wind speed and magnetic field strength in the pre-shock region, and variations in the IMF direction that do not modify significantly the shape of the intensity-time profiles; hence, the SaP model can be applied to describe this SEP event. First, the shock propagation is modelled to obtain the position of the particle source, and next the simulation of the transport of particles up to the observer's position (in this case, located at L1) is performed. For this, particles are assumed to be injected by the shock at the points in the shock front connected with the spacecraft through a Parker IMF line (i.e., at the cobpoints; Heras et al. 1995). In the SaP model, Eq. (4.11) is solved with the inclusion of the solar wind effects on the low-energy protons (Ruffolo 1995; Lario et al. 1998).

From the particle transport models, the evolution of the injection rate, Q and the proton mean free path, $\lambda_{\|}$ can be obtained by fitting the observed time profiles of particle (omnidirectional) intensities[1] and of the first order anisotropies (when available, e.g., Lario et al. 1998) or by fitting directly directional intensities (e.g., Agueda et al. 2008). For the April 2000 event, the omnidirectional intensity-time profiles are simultaneously fitted for eighteen energy channels and the first order anisotropies for $E < 2$ MeV, to better constrain the values of the parameters used (see details in e.g., Pomoell et al. 2015).

The resulting values of $\lambda_{\|}$ at 9.1 MeV are: 1.30 AU for $t < 11.0$ h, 0.65 AU for $11.0 \leq t < 15.8$ h and 0.33 AU for $t \geq 15.8$ h, assuming $\bar{q} = 1.6$ for its rigidity dependence. Figure 4.4 shows the derived evolution of the source function, Q, that continuously changes from low to high energies. For $E > 36.4$ MeV, Q decreases rapidly (two orders of magnitude in 10 h). After this time, the shock, located already at 80 $R_{\odot}$ is no longer efficiently injecting >40 MeV protons. For lower energies, Q continuously decreases, this decrease being slow for lower energies. The use of a particle transport model (in this case the SaP model) yields the quantitative description of how the shock is gradually losing efficiency at injecting particles as it moves away from the Sun and as the magnetic connection with the observer varies.

[1]Hereafter 'intensity' is used for 'differential intensity'.

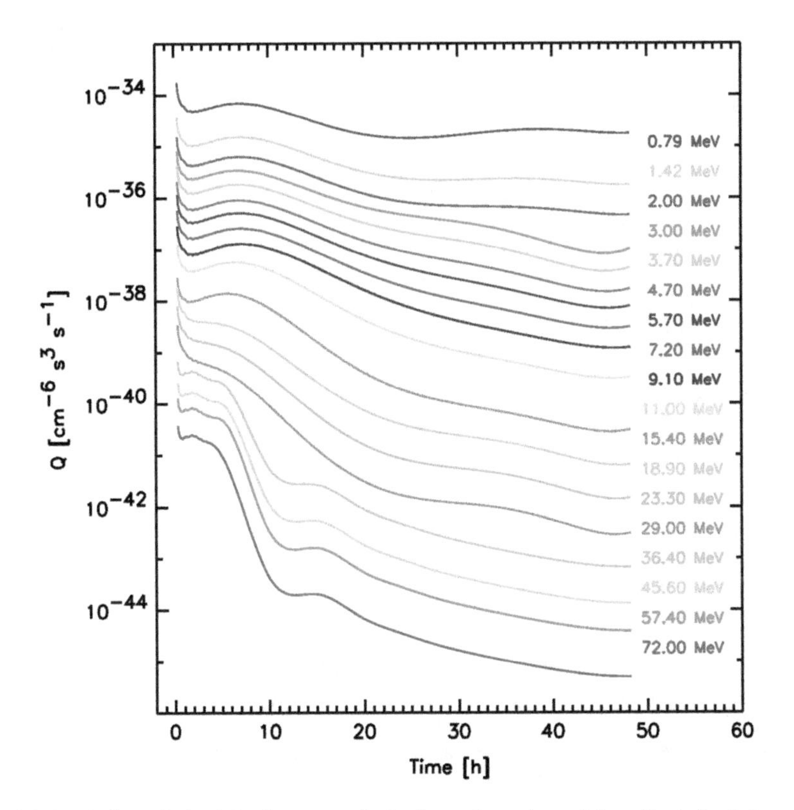

Fig. 4.4 Evolution of the injection rate of shock-accelerated particles, Q, derived from the modelling of the SEP event with the SaP model. Each curve, *colour coded* as indicated in the inset, corresponds to the injection rate profile derived for each of the eighteen energy channels modelled and computed from the first cobpoint up to the shock arrival at the ACE spacecraft

4.3.3 Inferring Transport Conditions

An important aspect to consider when inferring the transport conditions is the level of freedom in performing the fitting of observed intensities. If Q and λ_r values are derived employing only the omnidirectional intensity profile, for a given energy channel, the problem is ill-posed. Additional information is needed and can be extracted from first order anisotropies or from directional intensities. To illustrate this point the inversion model by Agueda et al. (2008, 2012) is utilised. This model assumes a fixed source of near-relativistic electrons placed at $2\,R_{\odot}$ from the Sun. The middle column of Fig. 4.5 shows an omnidirectional intensity-time profile (red curve) that is fitted with the model (black curves). In the left column are shown four different electron injection histories, $Q(t)$, and values of λ_r resulting from the fitting. The injection profile and λ_r shown in the first row do not fit the data, as it is clearly seen in the middle panel; however, the remainder three combinations shown in the next rows do fit the omnidirectional intensities perfectly; thus, indicating that

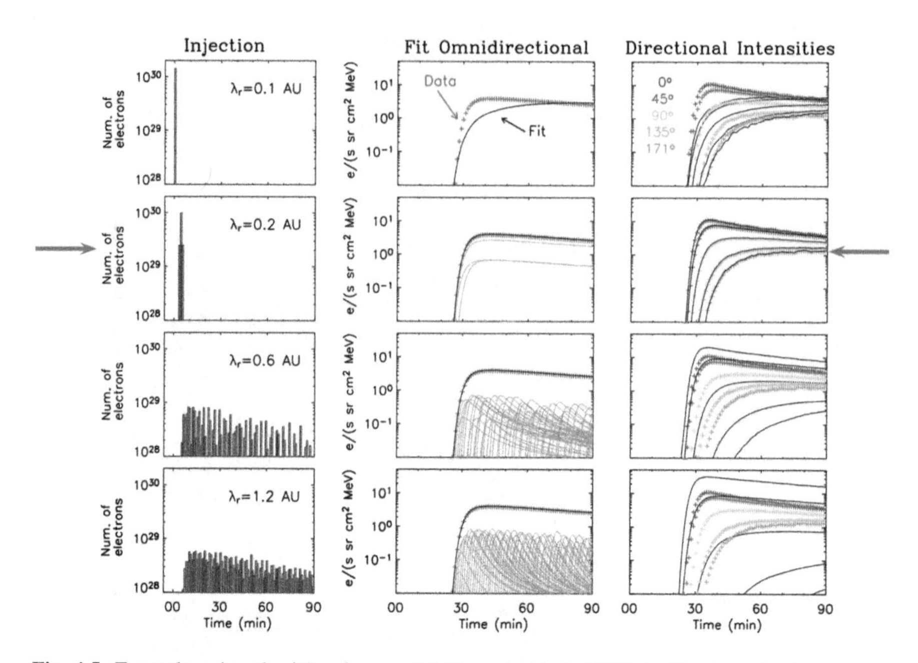

Fig. 4.5 Example using the inversion model (Agueda et al. 2008) to illustrate the importance of having directional information of particle intensities. Near-relativistic electron injection history profiles and λ_r (*left column*), omnidirectional intensities (*middle column*) and directional intensities (*right column*) are shown for different scenarios. The model fits (*black curves*) reproduce the observations (*colour curves*) only in the second case. See text for details

more information is needed to find the correct solution. In the right panel of Fig. 4.5 the corresponding directional intensities (colour curves) are shown. The model fits (black curves) only reproduce the directional intensities correctly in the second case; thus showing that directional intensities are crucial for inferring the correct parameters. Directional intensities or angular information of the particle distribution function is not always provided by instrumentation onboard spacecraft. It is usually available for near-relativistic electrons and low-energy protons, but not for high-energy protons for which measurement statistics are relatively low.

4.4 Concluding Remarks

In this chapter the key basic facts of the interplanetary scenario and of the main effects to consider when simulating the SEP transport in interplanetary space have been presented. The interested reader may follow the various references provided to deepen in the study of SEP transport. In summary, the main messages to take away are: the average interplanetary magnetic field can be described by an Archimedean spiral. Superposed on this average spiral field are small-scale random fluctuations. In the co-rotating frame, the motion of the guiding centre of SEPs travels along the mean spiral direction and particles are scattered by small-scale fluctuations embedded in the solar wind. Interplanetary transport models help to infer the particle source function of SEP events.

References

Afanasiev, A., Vainio, R.: Astrophys. J. Suppl. **207**, 29 (2013). doi:10.1088/0067-0049/207/2/29
Agueda, N., Vainio, R.: J. Space Weather Space Clim. **3**(27), A10 (2013). doi:10.1051/swsc/2013034
Agueda, N., Vainio, R., Lario, D., Sanahuja, B.: Astrophys. J. **675**, 1601–1613 (2008). doi:10.1086/527527
Agueda, N., Vainio, R., Sanahuja, B.: Astrophys. J. Suppl. **202**, 18 (2012). doi:10.1088/0067-0049/202/2/18
Bieber, J.W., Burger, R.A., Matthaeus, W.H.: Int. Cosmic Ray Conf. **4**, 694 (1995)
Bittencourt, J.A.: Fundamentals of Plasma Physics, 3rd edn. Springer, New York (2004)
Dröge, W.: Astrophys. J. **537**, 1073 (2000). doi: 10.1086/309080
Gold, R.E., Krimigis, S.M., Hawkins, S.E., Haggerty, III, D.K., Lohr, D.A., Fiore, E., Armstrong, T.P., Holland, G., Lanzerotti, L.J.: Space Sci. Rev. **86**, 541 (1998). doi:10.1023/A:1005088115759
Hasselmann, K., Wibberenz, G.: Astrophys. J. **162**, 1049 (1970). doi:10.1086/150736
Heras, A.M., Sanahuja, B., Lario, D., Smith, Z.K., Detman, T., Dryer, M.: Astrophys. J. **445**, 497 (1995). doi:10.1086/175714
Hundhausen, A.J.: In: Kivelson, M.G., Russell C.T. (eds.) Introduction to Space Physics, pp. 91–128. Cambridge University Press, New York (1995), Chap. 4
Jokipii, J.R.: Rev. Geophys. Space Phys. **9**, 27 (1971). doi:10.1029/RG009i001p00027
Jokipii, J.R.: Astrophys. J. **313**, 842 (1987). doi:10.1086/165022
Kallenrode, M.B.: Space Physics: An Introduction to Plasmas and Particles in the Heliosphere and Magnetospheres. Springer, Berlin (2004)
Kunow, H., Wibberenz, G., Green, G., Müller-Mellin, R., Kallenrode, M.B.: Energetic particles in the inner solar system. In: Schwenn, R., Marsch, E. (eds.) Physics of the Inner Heliosphere II. Physics and Chemistry in Space (Space and Solar Physics), vol. 21, Springer, Berlin, Heidelberg (1991). https://doi.org/10.1007/978-3-642-75364-0_6, https://link.springer.com/chapter/10.1007/978-3-642-75364-0_6
Lario, D., Sanahuja, B., Heras, A.M.: Astrophys. J. **509**, 415 (1998). doi:10.1086/306461
McComas, D.J., Bame, S.J., Barker, P., Feldman, W.C., Phillips, J.L., Riley, P., Griffee, J.W.: Space Sci. Rev. **86**, 563 (1998). doi:10.1023/A:1005040232597
Parker, E.N.: Astrophys. J. **128**, 664 (1958). doi:10.1086/146579
Parker, E.N.: Planet. Space Sci. **13**, 9 (1965). doi:10.1016/0032-0633(65)90131-5
Pomoell, J., Aran, A., Jacobs, C., Rodríguez-Gasén, R., Poedts, S., Sanahuja, B.: J. Space Weather Space Clim. **5**(27), A12 (2015). doi:10.1051/swsc/2015015
Roelof, E.C.: In: Ögelman, H., Wayland J.R. (eds.) Lectures in High-Energy Astrophysics, p. 111. NASA, Washington DC (1969)
Ruffolo, D.: Astrophys. J. **442**, 861 (1995). doi:10.1086/175489
Smith, E.J.: In: Balog, A., Lanzerotti, L.J., Suess S.T. (eds.) The Heliosphere through the Solar Activity Cycle, pp. 79–150. Springer-Praxis, Printed in Germany (2008), Chap. 4
Smith, C.W., L'Heureux, J., Ness, N.F., Acuña, M.H., Burlaga, L.F., Scheifele, J.: Space Sci. Rev. **86**, 613 (1998). doi:10.1023/A:1005092216668

Tan, L.C., Reames, D.V., Ng, C.K., Saloniemi, O., Wang, L.: Astrophys. J. **701**, 1753 (2009). doi:10.1086/0004-637X/701/2/1753

Torsti, J., Valtonen, E., Lumme, M., Peltonen, P., Eronen, T., Louhola, M., Riihonen, E., Schultz, G., Teittinen, M., Ahola, K., Holmlund, C., Kelhä, V., Leppälä, K., Ruuska, P., Strömmer, E.: Sol. Phys. **162**, 505 (1995). doi:10.1007/BF00733438

Zank, G.P.: Transport Processes in Space Physics and Astrophysics. Lecture Notes in Physics, vol. 877. Springer, New York (2014)

Chapter 5
Cosmic Ray Particle Transport in the Earth's Magnetosphere

R. Bütikofer

Abstract The transport of the cosmic ray particles in the Earth's magnetic field must be considered for cosmic ray investigations based on cosmic ray measurements in the geomagnetosphere. The motion of charged particles in a magnetic field is defined by the Lorentz force. The trajectories of cosmic ray particles are curved by the Earth's magnetic field. In a first approximation the geomagnetic field can be described by a dipole magnetic field. For a more accurate description the geomagnetic magnetic field is divided into two parts: the inner part generated by an internal dynamo and the outer part induced by different current systems in the ionosphere and the magnetosphere. Models have been developed that describe the inner and the outer magnetic field. The computations of the propagation of the cosmic ray particles in the Earth's magnetosphere are made with computer programs based on numerical integration of the equation of motion. For the specification of geomagnetic effects on cosmic ray particles the concept of cutoff rigidities and of asymptotic directions have been introduced.

5.1 Introduction

Charged particles moving in a magnetic field are deflected. Investigations of cosmic ray observations based on ground-based and on space-based detectors inside the geomagnetosphere require therefore a detailed knowledge of the propagation of the cosmic ray particles in the Earth's magnetic field. The conditions under which a cosmic ray particle has access to a specific point of observation are defined by the Earth's magnetic field, the energy of the particle as well as the direction of incidence.

The implementation of the quantity "magnetic rigidity" is useful as particles with the same rigidity R, charge sign and initial conditions have identical trajectories in a

R. Bütikofer (✉)
University of Bern, Physikalisches Institut, Sidlerstrasse 5, CH-3012 Bern, Switzerland

High Altitude Research Stations Jungfraujoch and Gornergrat, Sidlerstrasse 5, CH-3012 Bern, Switzerland
e-mail: rolf.buetikofer@space.unibe.ch

O.E. Malandraki, N.B. Crosby (eds.), *Solar Particle Radiation Storms Forecasting and Analysis, The HESPERIA HORIZON 2020 Project and Beyond*, Astrophysics and Space Science Library 444, DOI 10.1007/978-3-319-60051-2_5

static magnetic field. The rigidity R is defined as $R = \frac{pc}{Ze}$, where p is the momentum, c is the speed of light and Ze is the charge of the cosmic ray particle. The unit of the rigidity R is volts. A convenient unit is GV (10^9 V).
History:

~1st cent. AD	First known magnetic compass was invented in China
13th cent. AD	First theories about geomagnetism
In early 1890s	Hendrik Lorentz derived the equation that describes the forces on a charged particle in an electromagnetic field, the so-called Lorentz force
1912	Discovery of the cosmic ray by the Austrian Victor Hess
1930	First works were started by Störmer to understand the geomagnetic effects on cosmic rays (Störmer 1930)
1957/58	International Geophysical Year Systematic investigations of the effects of the Earth's magnetic field on the cosmic rays as observed on the ground

For a more detailed historical overview, see e.g. Smart et al. (2000).
The forces that act on moving charged particles in a electromagnetic field and their effects on the particles' trajectories are described in Sect. 5.2. The Earth's magnetic field, i.e. the inner and the outer part of the Earth's magnetic field as well as the models describing these fields, is addressed in Sect. 5.3. In Sect. 5.4 the differential equations that describe the motion of the charged cosmic ray particles in the Earth's magnetic field and their numerical computation are summarised. The concepts of "cutoff rigidities" and of "asymptotic directions" that have been introduced to quantify the geomagnetic field effect are described in Sect. 5.5.

5.2 Motion of Charged Particles in a Magnetic Field: Lorentz Force

The combination of electric and magnetic forces on a charged particle due to electromagnetic fields is described by the Lorentz force **F**:

$$\mathbf{F} = Ze\,\mathbf{E} + Ze\,[\mathbf{v} \times \mathbf{B}] \tag{5.1}$$

where

Ze	charge of the moving particle, e is the elementary charge
E	electric field
v	velocity of the particle
B	magnetic field

The effect of electric fields can be neglected in the geomagnetosphere due to its high electric conductivity.

The force equation becomes:

$$\mathbf{F} = m \cdot \frac{d\mathbf{v}}{dt} = Ze\,[\mathbf{v} \times \mathbf{B}] \tag{5.2}$$

A charged particle is accelerated perpendicularly to the speed $\mathbf{v}$, it follows that the absolute value of the momentum $|m \cdot \mathbf{v}|$ and its kinetic energy are conserved.
For relativistic particles with mass $m = \gamma m_0$, where γ is the Lorentz factor ($\gamma = (1 - v^2/c^2)^{-\frac{1}{2}}$) and m_0 is the rest mass of the particle, it follows:

$$\frac{d\mathbf{v}}{dt} = -\frac{Ze}{\gamma m_0}\,[\mathbf{v} \times \mathbf{B}] \tag{5.3}$$

For a moving charged particle in a uniform magnetic field, the speed vector $\mathbf{v}$ can be split into a component parallel $v_{\parallel}$ and perpendicular $v_{\perp}$ to the magnetic field $\mathbf{B}$. The motion of the particle is then described by a movement with constant speed along the magnetic field $v_{\parallel}$ and a circular motion around the magnetic field lines.
The centripetal acceleration is

$$\frac{{v_{\perp}}^2}{r_c} = \frac{Ze}{\gamma m_0} \cdot v_{\perp} \cdot |\mathbf{B}| \tag{5.4}$$

where r_c is the cyclotron radius or gyroradius.
For the cyclotron radius r_c follows:

$$r_c = \frac{\gamma m_0 \cdot v_{\perp}}{Ze \cdot |\mathbf{B}|} \tag{5.5}$$

The above formula for the cyclotron radius can be rearranged to give a more practical expression for an estimation of the cosmic ray trajectory characteristic:

$$r_c\,[\text{meter}] \;=\; 3.3 \times \frac{(\gamma m_0 c^2\,[\text{GeV}]) \,\cdot\, (v_{\perp}\,/\,c)}{(Z \,\cdot\, |\mathbf{B}|\,[\text{Tesla}])} \tag{5.6}$$

where GeV is the unit of Giga-electronVolts.

A proton with kinetic energy $E_{kin} = 10\,\text{GeV}$ in the Earth's magnetic field close to the Earth where $|\mathbf{B}| \approx 30{,}000\,\text{nT}$ (Sect. 5.3.1) and $\mathbf{v} \perp \mathbf{B}$, i.e. $v_{\perp} = 0.99\,c$, has a gyroradius in the order of 10^6 m or 0.15 Earth radii.

5.3 Earth's Magnetic Field

The geomagnetosphere is the region close to the Earth where the motion of charged particles is mainly determined by the Earth's magnetic field. The size, the shape and the inner structure of the geomagnetosphere is configured by the interaction of the solar wind with the Earth's magnetic field. The extension of the geomagnetosphere in space is therefore determined by the equilibrium between the pressure of the streaming solar wind plasma and the magnetic pressure of the magnetosphere. The front end of the magnetopause is at a standoff distance of 10–12 Earth's radii (Re) from the Earth's centre during quiescent solar wind conditions. The magnetotail has a length of at least 100 Re. During times of disturbed solar wind conditions the characteristics of the geomagnetosphere are changed and are defined by the solar wind speed, the particle density in the solar wind, the strength and the direction of the interplanetary magnetic field. In addition, the relative position of the magnetic dipole inside the Earth defines the characteristics of the geomagnetic field. In the models of the inner Earth's magnetic field these effects must be considered.

The magnetosheath, the space between the magnetopause and the bow shock, is the consequence of the fact that the solar wind can not penetrate the Earth's magnetic field because of its high electric conductivity and that therefore the magnetic field of the solar wind must be swept around the magnetopause. The magnetic field strength in the magnetosheath may be in the order of a few 10 nT (Kobel and Fluckiger 1994). The geomagnetic field line topology in the geomagnetosphere is illustrated in Fig. 5.1.

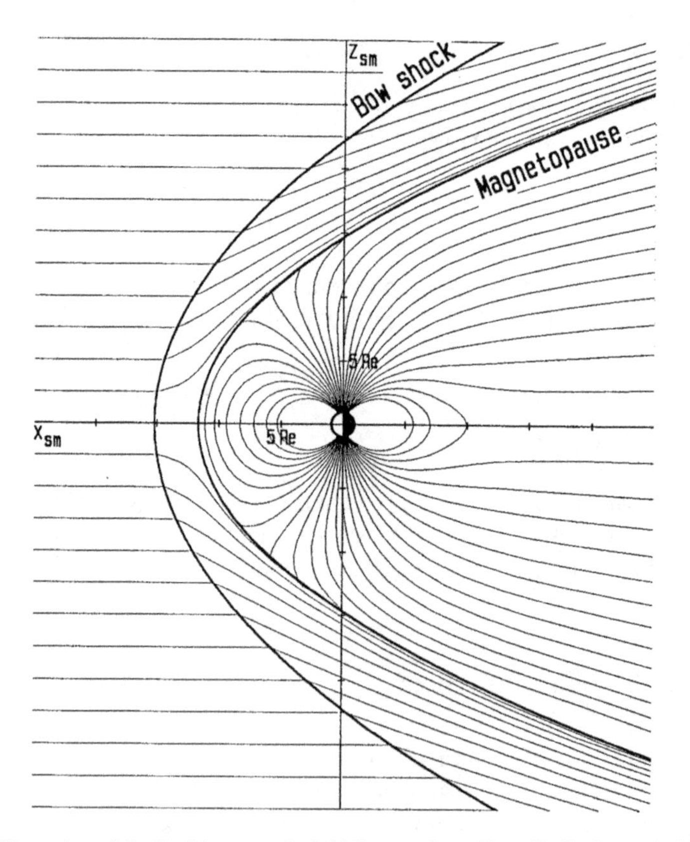

Fig. 5.1 Illustration of the Earth's magnetic field line topology. From Bütikofer et al. (1995)

Close to the Earth the magnetic field can be described in a first approximation by a geocentric dipole (Sect. 5.3.1). In a more accurate description the Earth's magnetic field is divided into an inner and an outer part. The inner magnetic field is produced by sources in the interior of the Earth. It is described by the IGRF model which is based on magnetic field measurements at the Earth's surface (Sect. 5.3.2). The external magnetic field is produced by the different electric current systems in the ionosphere and the inner magnetosphere. There exist different field models that describe the external Earth's magnetic field. The models by Tsyganenko—the models that are mainly used for cosmic ray trajectory computations within the geomagnetosphere—are presented in Sect. 5.3.4.

5.3.1 The Magnetic Field of the Earth as a Dipole Field

In a first approximation the Earth's magnetic field can be described by a geocentric magnetic dipole.

The magnetic moment **m** of a circular current is:

$$|\mathbf{m}| = I \cdot S \tag{5.7}$$

where

I electric current
S area that is spanned by the circular current

For the Earth dipole: $|\mathbf{m}| = 8.1 \cdot 10^{22}\ \mathrm{A\,m^2}$
The geomagnetic moment is defined as:

$$\mathbf{M} = \frac{\mu_0}{4\pi}\,\mathbf{m} \tag{5.8}$$

where the vacuum permeability $\mu_0 = 4\pi \cdot 10^{-7}\ \mathrm{Vs\,/\,Am}$

$$\begin{aligned} |\mathbf{M}| &= 8.1 \cdot 10^{25}\ \mathrm{Gauss\ cm^3} \\ &= 8.1 \cdot 10^{15}\ \mathrm{V\,s\,m} \end{aligned}$$

In polar coordinates the components of the magnetic field $\mathbf{B}(r, \lambda, \varphi)$ are:

$$\begin{aligned} B_r &= -\frac{2\,|\mathbf{M}|\,\sin\lambda}{r^3} \\ B_\lambda &= -\frac{|\mathbf{M}|\,\cos\lambda}{r^3} \\ B_\varphi &= 0 \end{aligned} \tag{5.9}$$

where

r distance from Earth's center
λ geomagnetic latitude
φ geomagnetic longitude

Magnetic field strength at the geomagnetic equator: $B_\lambda = \frac{M}{r^3}$, $B_r = 0$, i.e. at Earth's surface ($r = 6.4 \cdot 10^6$ m) $B_\lambda \approx 30{,}000$ nT.
Magnetic field strength at the poles: $B_\lambda = 0$, $B_r = \frac{2M}{r^3}$, i.e. at Earth's surface $B_r \approx 60{,}000$ nT.
The magnetic strength as function of λ and r is:

$$|\mathbf{B}| = \frac{|\mathbf{M}|}{r^3}\sqrt{1 + 3\sin\lambda^2} \sim \frac{1}{r^3} \tag{5.10}$$

Units:

$$1\,\text{Gauss}\,(G) = 10^{-4}\,\text{V s/m}^2 = 10^{-4}\,\text{Tesla (T)}$$
$$1\,\text{Gamma}\,(\gamma) = 1\,\text{nT}$$

5.3.2 *Magnetic Field Model Due to Internal Sources: IGRF*

The main geomagnetic field is generated primarily by a hydrodynamic geodynamo in the Earth's fluid outer core which varies slowly with time. The geodynamo has an underlying offset dipolar configuration which is currently tilted at an angle of about 10° with respect to the Earth's rotational axis. The time dependence in the magnetic field models is usually approached by a sequence of static configurations. For a static magnetic field the equations of the magnetostatics must be fulfilled as a special case of the Maxwell's equations:

$$\nabla \cdot \mathbf{B} = 0$$
$$\nabla \times \mathbf{B} = \mu_0\,\mathbf{j} \qquad \mathbf{j}\text{: current density}$$

There are no currents to the center of the Earth:

$$\mathbf{j} = 0, \text{ i.e. } \nabla \times \mathbf{B} = 0$$

From this it follows that there exists a potential V, so that

$$\mathbf{B} = -\nabla V$$

Since the magnetic field **B** is divergence-free (Gauss's law for magnetism) it follows for $r \geq \mathrm{Re}$:

$$\nabla^2 V = 0 \text{ or } \Delta V = 0 \quad \text{(Laplace's equation)}$$

Assumption for the solution of the Laplace equation: only sources from the Earth's interior are considered.

The geomagnetic main field is usually described by spherical harmonics (Chapman and Bartels 1940). Here, the representation of the geomagnetic main field with spherical harmonics according to the method by Gauss with the Schmidt normalization (Chapman and Bartels 1940) is used. In geographic spherical coordinates (r, θ, ϕ) the corresponding geomagnetic potential V can be expressed:

$$V(r,\theta,\phi) = a\sum_{n=1}^{\infty}\sum_{m=0}^{n}\left(\frac{a}{r}\right)^{n+1} \cdot P_n^m(\cos\theta) \cdot \{g_n^m \cos(m\phi) + h_n^m \sin(m\phi)\} \quad (5.11)$$

where

a mean Earth's radius, $a = 6371.2$ km

P_n^m Schmidt normalized (Chapman and Bartels 1940) associated Legendre functions of degree n and of order m

g_n^m, h_n^m Gauss coefficients

As a consequence of the secular variations in the geomagnetic field, the Gauss coefficients must be determined periodically. The variation of the declination is ~0.13°/year, the westward drift of the non-dipolar terms has a time period of ~2000 years, and the change in the dipole moment a period of some 1000 years (Merrill and McElhinny 1983). The International Union of Geodesy and Geophysics (IUGG) and the International Association of Geomagnetism and Aeronomy (IAGA) determine from measurements of the magnetic field **B** on the ground and publish every 5 years the Gauss coefficients g_n^m and h_n^m (Thébault et al. 2015). Before the year 2000, the parameters until degree $n = m = 10$ were determined and since the year 2000 until degree $n = m = 13$, see https://www.ngdc.noaa.gov/IAGA/vmod/igrf.html.
The components of $\mathbf{B}(r, \theta, \phi)$ in spherical coordinates are:

$$\begin{aligned} \mathbf{B}_r(r,\theta,\phi) &= -\frac{\partial V(r,\theta,\phi)}{\partial r} \\ \mathbf{B}_{\theta}(r,\theta,\phi) &= -\frac{1}{r}\frac{\partial V(r,\theta,\phi)}{\partial \theta} \\ \mathbf{B}_{\phi}(r,\theta,\phi) &= -\frac{1}{r\,sin\theta}\frac{\partial V(r,\theta,\phi)}{\partial \Phi} \end{aligned} \quad (5.12)$$

where

- r distance from Earth's center
- θ geographic co-latitude, $\theta = 90° - \Lambda$
- Λ geographic latitude
- ϕ geographic length

The strength of the magnetic field at the Earth's surface ranges from less than 30,000 nT or 0.3 Gauss in South America (South Atlantic Anomaly) to over 60,000 nT around the magnetic poles (northern Canada, Siberia and coast of Antarctica, south of Australia). For comparison, the strength of the interplanetary magnetic field near Earth is typically 5 nT, i.e. the Earth's magnetic field at the Earth surface is about four orders of magnitude larger.

5.3.3 *Contributions to the Earth's Magnetic Field by Magnetospheric Electric Currents*

In addition to the internal sources of the Earth's magnetic field, there is also a contribution of external origin: the electrical current systems in the ionosphere and the magnetosphere. Figure 5.2 shows a schematic view of the different current

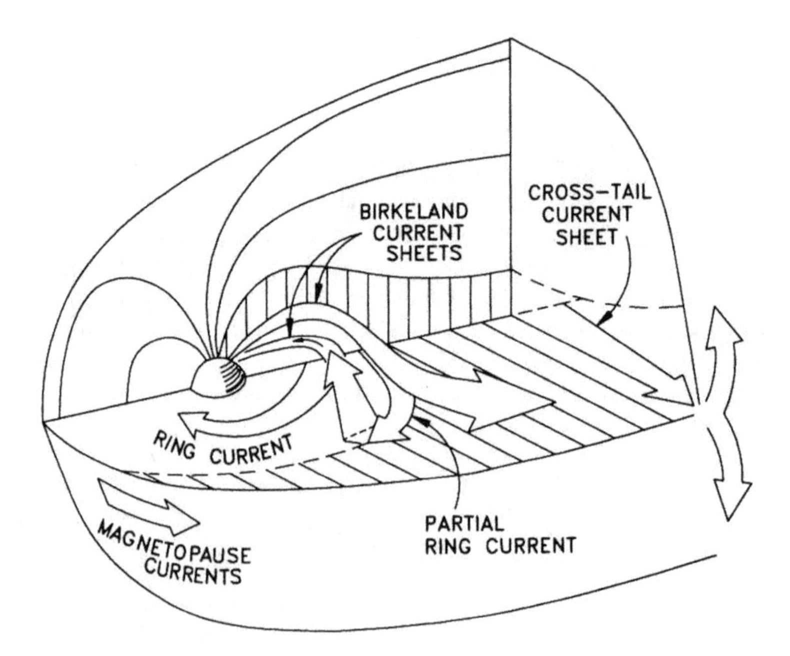

Fig. 5.2 Schematic view of the different current systems which contribute to the Earth's magnetic field.
(From Stern 1994, reproduced with permission from publisher John Wiley and Sons for electronic and press publishing, license number: 4118701268653, license date: May 30, 2017)

systems in the magnetosphere of the Earth. These current systems may vary rapidly, depending on the solar activity. During quiet periods the amplitudes of these external contributions are ~20 nT at mid-latitudes and may increase to more than the ten-fold during geomagnetic storms. The most important current systems are: the ring current in the radiation belts, the Chapman-Ferraro current on the magnetopause (magnetopause current), the field aligned (Birkeland) currents along geomagnetic field lines connecting the Earth's magnetosphere to the Earth's high latitude ionosphere, and the tail currents in the tail of the magnetosphere. The intensities of these currents reach millions of amperes and are related to the solar activity. During times of low solar activity, the standoff distance of the magnetopause currents is at ~10.5 Earth radii and the generated magnetic field close to the Earth is ~25 nT. The ring current, i.e. the longitudinal drift of energetic (10–200 keV) particles that are bouncing along the magnetic field lines between North and South polar regions, has a radius of ~6 R_E during quiet times and its contribution to the magnetic field at the Earth is about ~40 nT.

5.3.4 Magnetic Field Models of the External Sources

Since the 1970s efforts were made to improve the quantitative quality of the magnetospheric magnetic field models, see e.g. the review paper by Walker (1979). Most of the models include currents in the inner magnetosphere in addition to the boundary currents and the magnetotail current system. All the models include the tilt angle of the internal magnetic dipole as an input parameter. With the advent of the space era it became possible to extend the models from low to high altitudes, eventually including even the entire magnetosphere. However, the modeling of the magnetic field in that region is much more difficult, mostly because the magnetic field from external sources (currents in the geomagnetosphere) predominates the magnetic field with growing distance from the Earth. Vector measurements of the magnetic field should be made throughout the entire space where the field should be modeled, i.e. it is necessary to accumulate large amounts of space magnetometer data taken in a wide range of the geomagnetosphere. In contrast to the main geomagnetic field (variations on a timescale of thousands of years), the magnetic field in the outer regions of the geomagnetosphere is a very dynamical system on short time scales and depends on different factors. The first factor is the orientation of the Earth's magnetic axis with respect to the direction of the incoming solar wind flow, which varies with time because of the Earth's diurnal rotation and its yearly orbital motion around the Sun, and the frequent variations of the solar wind characteristics. Another important factor is the state of the solar wind, in particular, the orientation and strength of the interplanetary magnetic field. The interaction between the terrestrial and the interplanetary magnetic fields becomes strongly effective when the interplanetary magnetic field is antiparallel to the Earth's field on the dayside boundary of the magnetosphere. In this case, the geomagnetic and the interplanetary field lines connect across the magnetospheric boundary, which strongly enhances the transfer of the solar wind mass, energy, and electric field inside the magnetosphere.

Different models have been developed to describe the magnetic field in the whole geomagnetosphere (Mead and Fairfield 1975; Olson and Pfitzer 1988; Tsyganenko 1987, 1989, 1996; Ostapenko and Maltsev 1997; Tsyganenko 2002a,b; Tsyganenko and Sitnov 2005). For the determination of cosmic ray particle trajectories mainly the magnetic field models by Tsyganenko (1989, 1996, 2002a,b), Ostapenko and Maltsev (1997) and Tsyganenko and Sitnov (2005) are used.

The Tsyganenko models are semi-empirical best-fit representations for the magnetic field, based on a large number of satellite observations (IMP, HEOS, ISEE, POLAR, Geotail, etc.). The models include the contributions from external magnetospheric sources: ring current, magnetotail current system, magnetopause currents and large-scale system of field-aligned currents.

Tsyganenko model T89 (Tsyganenko 1989) was primarily developed as a tail model. It is based on satellite measurements at distances from the Earth less than 70 R_E, therefore its domain of validity is limited to this region in space. It provides seven different states of the geomagnetosphere corresponding to different levels of the geomagnetic activity represented by the *Kp*-index[1] 0, 1, ..., ≥6. The model does not consider the continuous variation of the structure of the magnetosphere as a function of geomagnetic indices like *Dst* and of solar wind parameters. The consideration of these parameters to describe the evolution of the magnetosphere is in particular important during a magnetic storm. Therefore the use of T89 is not reasonable during times when the geomagnetic field is strongly disturbed.

Tsyganenko model T96 (Tsyganenko 1996) considers in contrast to the T89 model the continuous variation of the structure of the magnetosphere as a function of the geomagnetic indices like *Dst* and of the solar wind parameters. In this model the external magnetospheric magnetic field is generated by different current systems where the shape and the strength depend on the dipole tilt angle, on the solar wind dynamic pressure, on the *Dst* index and on the interplanetary magnetic field components B_y^{GSM} and B_z^{GSM} in geocentric solar magnetospheric coordinates (GSM). This model has an explicitly defined realistic magnetopause which is represented by a semi ellipsoid of rotation towards the Sun and by a cylindrical surface in the far tail for $x^{GSM} \leq -60 R_E$.

Tsyganenko model T01 (Tsyganenko 2002a,b) is based on the same principles as the model T96 but has essential improvements. The T01 model considers the variable configuration of the inner and near magnetosphere for different interplanetary conditions and ground disturbance levels.

[1]The *Kp*-index is a quasi-logarithmic quantity for the variation of the magnetic field intensity at the Earth's surface as function of time. The range of the *Kp*-index is 0° (for quite conditions) over 0+, 1−, 1+ to 9° (for extreme disturbed magnetic field (geomagnetic storm)). It is derived from the maximum fluctuations of the horizontal components of the Earth's magnetic field observed by observation stations around the world and is published every 3 h.

Tsyganenko model T04 (Tsyganenko and Sitnov 2005) is a dynamical model of the storm-time geomagnetic field in the inner magnetosphere, using space magnetometer data taken during 37 major events in 1996–2000 and concurrent observations of the solar wind and the interplanetary magnetic field. Therefore, this model is only applicable for times with strong disturbances of the geomagnetic field.

The Tsyganenko model T89 is usually used to compute cosmic ray trajectories in the geomagnetosphere due to the much simpler utilisation with only a few input parameters (date, time, *Kp*-index) and the much less time-consuming computation effort compared to the other Tsyganenko models.

The magnetic field of the magnetosheath is usually not considered in the computation of cosmic ray particle trajectories, although the change in the direction of approach at the border of the geomagnetosphere due to the effect of the magnetosheath may be a few 10° at low rigidities ($R \sim 1$GV) (Bütikofer et al. 1997).

5.4 Computation of the Propagation of Cosmic Ray Particles in the Earth's Magnetosphere

There exists no solution of the equation of motion of a charged particle in the geomagnetosphere magnetic field in a closed form. Therefore the determination of cosmic ray trajectories in the geomagnetosphere is almost exclusively made by numerical integration on computer by using a model of the magnetic field in the geomagnetosphere. The cosmic ray particle trajectories are computed backward, i.e. starting at the location of observation and compute the trajectory away from the Earth. Thereby the effect is used that the path of a negatively charged particle with mass, m, charge, Ze, and speed, $\mathbf{v}$, in a static magnetic field, $\mathbf{B}$, is identical to that of an identical but positively charged particle with reverse sign of the velocity vector. For observation locations at ground the computations start at an altitude of typically 20 km above sea level as the interactions of primary cosmic ray particles with atoms and atomic nuclei in the atmosphere become important below this altitude (Smart et al. 2000).

The equations of motion of charged particles in a known magnetic field $\mathbf{B}(r, \theta, \phi)$ in a spherical coordinate system (r, θ, ϕ) are:

$$
\begin{aligned}
\frac{\mathrm{d}v_r}{\mathrm{d}t} &= \frac{Ze}{m}(v_\theta B_\phi - v_\phi B_\theta) + \frac{v_\theta^2}{r} + \frac{v_\phi^2}{r} \\
\frac{\mathrm{d}v_\theta}{\mathrm{d}t} &= \frac{Ze}{m}(v_\phi B_r - v_r B_\phi) - \frac{v_r v_\theta}{r} + \frac{v_\phi^2}{r\tan\theta} \\
\frac{\mathrm{d}v_\phi}{\mathrm{d}t} &= \frac{Ze}{m}(v_r B_\theta - v_\theta B_r) - \frac{v_r v_\phi}{r} - \frac{v_\theta v_\phi}{r\tan\theta}
\end{aligned}
\tag{5.13}
$$

$$\frac{\mathrm{d}r}{\mathrm{d}t} = v_r$$

$$\frac{\mathrm{d}\theta}{\mathrm{d}t} = \frac{v_\theta}{r}$$

$$\frac{\mathrm{d}\phi}{\mathrm{d}t} = \frac{v_\phi}{r\sin\theta}$$

where B_r, B_θ, B_ϕ are the known magnetic field components, v_r, v_θ, v_ϕ are the particle velocity components, c is the speed of light, Ze and m are respectively the charge and the mass of the particle, and r is the radial distance of the location of the particle from the center of the Earth.

The statement of the problem of the particle trajectory computation belongs to the category of initial value problems. They start at a selected time t_0 and with a set of known variables r_0, θ_0, ϕ_0, v_{r_0}, v_{θ_0}, v_{ϕ_0}. From this set of initial values the corresponding values after a short time interval Δt, i.e. at time $t_0 + \Delta t$, can be computed.

There exist different types of numerical methods to solve the initial value problems e.g. Runge–Kutta or Bulirsch-Stoer method (Press et al. 1986). These methods optimize in different ways the step size of the numerical integration, i.e. the interval size Δt, on the one hand to prevent that the error per step exceeds a preset maximum value and on the other hand to reduce the computation time. Different computer codes for the cosmic ray trajectory computations in the Earth's magnetic field have been developed (see e.g. Shea and Smart 1975; Flueckiger and Kobel 1990; Desorgher et al. 2006).

Figure 5.3 shows an illustration of charged particle trajectories with different rigidities entering the Earth at the same location from zenith direction. The cosmic

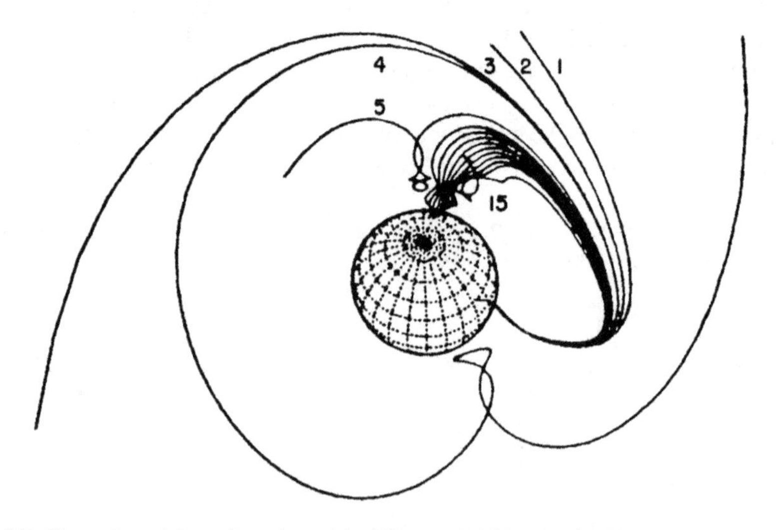

Fig. 5.3 Charged particle trajectories with different rigidities in the Earth's magnetic field. (From Smart et al. 2000, reproduced with permission from publisher Springer for electronic and press publishing, license number: 4118710609018, license date: May 30, 2017)

ray particle trajectories labeled 1–3 have high rigidities and are therefore less bent compared to the trajectories labeled with values >3. The trajectories 4 and 5 show loops, but both can escape the geomagnetosphere, i.e. cosmic ray particles with these rigidities can reach the specified location on Earth from zenith direction ("allowed trajectories"). Particles with even lower rigidities are more bent and the trajectories of these particles penetrate the Earth (re-entrant trajectories), i.e. particles with these rigidities can not reach the location of observation from zenith direction from outside of the geomagnetosphere ("forbidden trajectories").

5.5 The Concept of Cutoff Rigidities and Asymptotic Directions

The "cutoff rigidities" and the "asymptotic directions" have been introduced to specify the geomagnetic effects on cosmic ray particles and to determine the cosmic ray particle spectral characteristics and the anisotropy near Earth but outside the geomagnetosphere from cosmic ray measurements at ground (neutron monitors, muon detectors) or by space based detectors within the geomagnetosphere.

The cutoff rigidity at a selected location and with a specific direction of incidence is defined as the rigidity below which the cosmic ray particles have no access to this location from the given direction of incidence, i.e. trajectories with rigidities larger than the cutoff are "allowed trajectories" whereas trajectories with rigidities below the cutoff rigidity are "forbidden trajectories". The asymptotic direction of cosmic ray particles is used as the particle's trajectory direction of approach at the boundary of the geomagnetosphere.

The cutoff rigidities are determined by trajectory calculations at discrete rigidity intervals starting from a value above the highest possible cutoff rigidity down below the lowest possible allowed trajectory. The trajectories over this rigidity range show different features: first discontinuity in asymptotic direction, first forbidden trajectory, then usually a range of allowed and forbidden trajectories (co-called cosmic ray penumbra), lowest allowed trajectory.
The following parameters are used to describe the cutoff rigidity (Cooke et al. 1991):

- main cutoff rigidity R_M or upper cutoff R_u: rigidity of the last allowed trajectory before the first forbidden. This cutoff rigidity is close to the first discontinuity rigidity R_1, $R_M \approx R_1$
- Störmer cutoff R_S or lower cutoff R_l: rigidity of the last allowed trajectory, i.e. trajectories of particles with rigidities $< R_S$ are forbidden
- R_c: effective cutoff rigidity which is between R_u and R_l taking into account the penumbra, see Eq. (5.15).

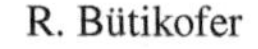

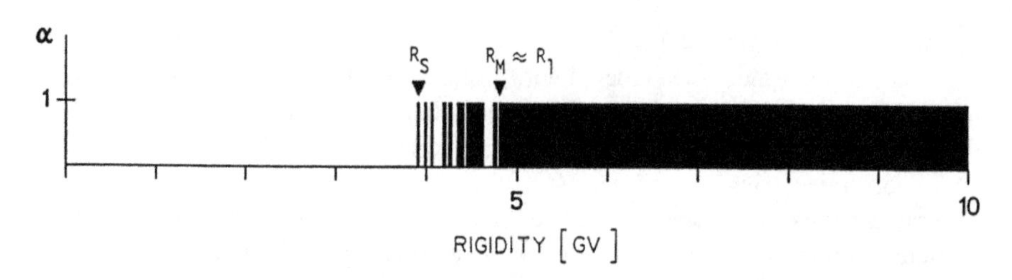

Fig. 5.4 Function $\alpha(R)$ for the station Jungfraujoch for vertical direction of incidence and corresponding cutoff rigidity values: R_S (Störmer cutoff), R_M (main cutoff), and R_1 (first-discontinuity rigidity)

For a location of observation and for a selected direction of incidence the effect of the Earth's magnetic field on the accessibility of cosmic ray particles is described by the filter function $\alpha(R)$:

$$\alpha(R) = \begin{cases} 0 & : \text{ if trajectory is forbidden for rigidity } R \\ 1 & : \text{ if trajectory is allowed for } R \end{cases} \tag{5.14}$$

Figure 5.4 shows the function $\alpha(R)$ for the station Jungfraujoch for vertical direction of incidence.

The effective cutoff rigidity R_c is given by

$$R_c = R_S + \int_{R_S}^{R_M} \alpha(R) dR \tag{5.15}$$

The effective geomagnetic cutoff rigidity R_c depends on the location of the observer, the direction of incidence into the atmosphere, the date and time, and the degree of disturbance of the geomagnetic field. The cutoff rigidities for ground-based cosmic ray stations and for vertical incidence range from $R_c \approx 0$ GV at the magnetic poles to $R_c \approx 15$ GV at the geomagnetic equator.

If one follows the cosmic ray particle's trajectory away from the Earth, the amount of bending per path length caused by the magnetic field is decreasing, i.e. the direction of the particle's trajectory approaches asymptotically its direction with no magnetic field. In the field of cosmic rays the expression asymptotic direction is used for the direction of the cosmic ray particle trajectory when it penetrates the border of the geomagnetosphere (magnetopause). The asymptotic direction of a cosmic ray particle that reaches the location of observation from a selected direction depends on the geographic coordinates of the observer and of the cosmic ray particle's rigidity. Figure 5.5 shows the trajectory of a cosmic ray particle reaching a location on the Earth from a selected direction and its puncture through the magnetopause. The arrow gives the direction of the trajectory at the puncture: the asymptotic direction.

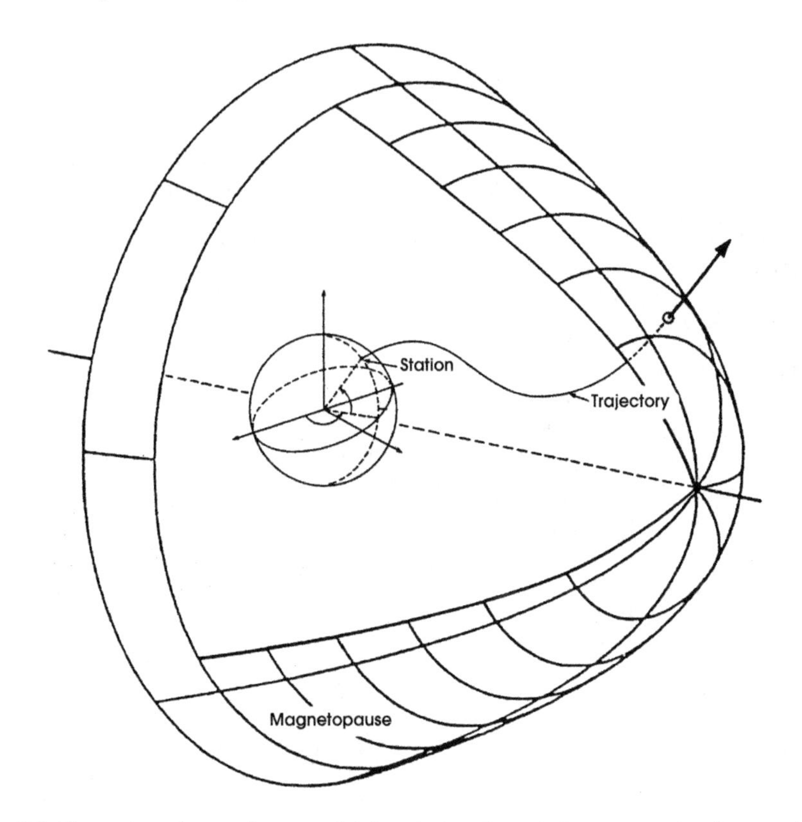

Fig. 5.5 Illustration of a cosmic ray particle's trajectory through the geomagnetosphere reaching a selected location on the Earth from a selected direction and of its related direction of approach at the magnetopause (asymptotic direction)

References

Bütikofer, R., Flückiger, E.O., Smart, D.F., Shea, M.A.: Effects of the magnetosheath on cosmic ray particle propagation in near-Earth space. Int. Cosmic Ray Conf. **4**, 1070 (1995)

Bütikofer, R., Flückiger, E.O., Smart, D.F., Shea, M.A.: Effects of the magnetosheath magnetic field on cosmic ray propagation near Earth. Int. Cosmic Ray Conf. **2**, 377 (1997)

Chapman, S., Bartels, J.: Geomagnetism. Oxford University Press, Oxford (1940)

Cooke, D.J., Humble, J.E., Shea, M.A., Smart, D.F., Lund, N.: On cosmic-ray cut-off terminology. Nuovo Cimento C Geophys. Space Phys. C **14**, 213–234 (1991)

Desorgher, L., Flückiger, E.O., Gurtner, M.: The PLANETOCOSMICS Geant4 application. In: 36th COSPAR Scientific Assembly. COSPAR Meeting, vol. 36, p. 2361 (2006)

Flueckiger, E.O., Kobel, E.: Aspects of combining models of the earth's internal and external magnetic field. J. Geomag. Geoelectr. **42**, 1123–1136 (1990)

Kobel, E., Fluckiger, E.O.: A model of the steady state magnetic field in the magnetosheath. J. Geophys. Res. **99**, 23 (1994)

Mead, G.D., Fairfield, D.H.: A quantitative magnetospheric model derived from spacecraft magnetometer data. J. Geophys. Res. **80**, 523–534 (1975)

Merrill, R.T., McElhinny, M.W.: The Earth's Magnetic Field: Its History, Origin and Planetary Perspective. International Geophysical Services, vol. 32. Academic Press, New York (1983)

Olson, W.P., Pfitzer, K.A.: Electric fields in earth orbital space. Technical report, Oct 1988

Ostapenko, A.A., Maltsev, Y.P.: Relation of the magnetic field in the magnetosphere to the geomagnetic and solar wind activity. J. Geophys. Res. **102**, 17467–17474 (1997)
Press, W.H., Flannery, B.P., Teukolsky, S.A.: Numerical Recipes. The Art of Scientific Computing. Cambridge University Press, Cambridge (1986)
Shea, M.A., Smart, D.F.: Asymptotic directions and vertical cutoff rigidities for selected cosmic-ray stations as calculated using the Finch and Leaton Geomagnetic field model. Technical report, April 1975
Smart, D.F., Shea, M.A., Flückiger, E.O.: Magnetospheric models and trajectory computations. Space Sci. Rev. **93**, 305–333 (2000)
Stern, D.P.: The art of mapping the magnetosphere. J. Geophys. Res. Space Phys. **99**(A9), 17169–17198 (1994)
Störmer, C.: Periodische Elektronenbahnen im Felde eines Elementarmagneten und ihre Anwendung auf Brüches Modellversuche und auf Eschenhagens Elementarwellen des Erdmagnetismus. Mit 32 Abbildungen. Zeits. Astrophys. **1**, 237 (1930)
Thébault, E., Finlay, C.C., Beggan, C.D., Alken, P., Aubert, J., Barrois, O., Bertrand, F., Bondar, T., Boness, A., Brocco, L., Canet, E., Chambodut, A., Chulliat, A., Coïsson, P., Civet, F., Du, A., Fournier, A., Fratter, I., Gillet, N., Hamilton, B., Hamoudi, M., Hulot, G., Jager, T., Korte, M., Kuang, W., Lalanne, X., Langlais, B., Léger, J.-M., Lesur, V., Lowes, F.J., Macmillan, S., Mandea, M., Manoj, C., Maus, S., Olsen, N., Petrov, V., Ridley, V., Rother, M., Sabaka, T.J., Saturnino, D., Schachtschneider, R., Sirol, O., Tangborn, A., Thomson, A., Tøffner-Clausen, L., Vigneron, P., Wardinski, I., Zvereva, T.: International geomagnetic reference field: the 12th generation. Earth Planets Space **67**, 79 (2015)
Tsyganenko, N.A.: Global quantitative models of the geomagnetic field in the cislunar magnetosphere for different disturbance levels. Planet. Space Sci. **35**, 1347–1358 (1987)
Tsyganenko, N.A.: A magnetospheric magnetic field model with a warped tail current sheet. Planet. Space Sci. **37**, 5–20 (1989)
Tsyganenko, N.A.: Effects of the solar wind conditions in the global magnetospheric configurations as deduced from data-based field models (Invited). In: Rolfe, E.J., Kaldeich, B. (eds.) International Conference on Substorms, vol. 389, p. 181. ESA Special Publication (1996)
Tsyganenko, N.A.: A model of the near magnetosphere with a dawn-dusk asymmetry 2. Parameterization and fitting to observations. J. Geophys. Res. (Space Phys.) **107**, 1176 (2002a)
Tsyganenko, N.A.: A model of the near magnetosphere with a dawn-dusk asymmetry 1. Mathematical structure. J. Geophys. Res. (Space Phys.) **107**, 1179 (2002b)
Tsyganenko, N.A., Sitnov, M.I.: Modeling the dynamics of the inner magnetosphere during strong geomagnetic storms. J. Geophys. Res. (Space Phys.) **110**, A03208 (2005)
Walker, R.J.: Quantitative modeling of planetary magnetospheric magnetic field. In: Quantitative Modeling of Magnetospheric Processes. American Geophysical Union, Washington DC (1979)

Chapter 6
Ground-Based Measurements of Energetic Particles by Neutron Monitors

R. Bütikofer

Abstract Since the International Geophysical Year (IGY) in 1957/58, the worldwide network of neutron monitors is the standard instrument to investigate the variations of the cosmic ray flux near Earth (11-year modulation of the galactic cosmic rays, Forbush decreases, solar cosmic ray events) in the GeV range. The ensemble of neutron monitors together with the geomagnetic field acts as a giant spectrometer and enables to deduce information about the primary cosmic ray spectrum near Earth in the energy range ~500 MeV to ~15 GeV. For the interpretation of the ground-based neutron monitor measurements, the transport of the cosmic rays in the Earth's magnetic field as well as the transport in the Earth's atmosphere and the detection efficiency of the secondary nucleons by the neutron monitors must be known. The Neutron Monitor Data Base (NMDB) developed in 2008/09 enables a rapid accessibility to the data of the worldwide neutron monitor network. A considerable number of neutron monitor stations send their data to NMDB in real-time which enables the operation of space weather applications based on neutron monitor data.

6.1 Introduction

The ground-based neutron monitors are relatively simple instruments in respect to technology and electronics. They are ideally suited to measure the intensity of the nucleonic component of the secondary cosmic radiation in the Earth's atmosphere and respond to primary cosmic ray particles in the GeV-range. Even after 60 years of operation the neutron monitors remain the state-of-the-art instrument for measuring the intensity variations of the primary cosmic rays in the energy range from ~500 MeV to ~30 GeV. This energy region complements the range above the energies covered by space-based cosmic ray detectors. The worldwide network of

R. Bütikofer (✉)
University of Bern, Physikalisches Institut, Sidlerstrasse 5, CH-3012 Bern, Switzerland

High Altitude Research Stations Jungfraujoch and Gornergrat, Sidlerstrasse 5, CH-3012 Bern, Switzerland
e-mail: rolf.buetikofer@space.unibe.ch

O.E. Malandraki, N.B. Crosby (eds.), *Solar Particle Radiation Storms Forecasting and Analysis, The HESPERIA HORIZON 2020 Project and Beyond*, Astrophysics and Space Science Library 444, DOI 10.1007/978-3-319-60051-2_6

neutron monitors is an excellent tool to investigate variations of the primary cosmic ray flux near Earth such as 11-year modulation and the sudden transient effects as Forbush decreases[1] and solar cosmic ray events. Since recently, neutron monitor measurements are also an important input for space weather applications.

In the International Geophysical Year 1957/58 the worldwide network of standardised neutron monitors was developed to investigate the variations of the cosmic ray intensity near Earth. There are two types of standardised neutron monitors in operation. The IGY (for International Geophysical Year) type was designed by Simpson (1955) in the early 1950s. About 10 years later Carmichael (1968) designed the larger NM64 monitor with an increased counting rate. Figure 6.1 shows the 18-IGY neutron monitor at Jungfraujoch, Switzerland (left) and the 6-NM64 monitor of Athens, Greece (right). The digits 18 respectively 6 give the number of the counter tubes deployed in the respective neutron monitor.

The ensemble of neutron monitors together with the geomagnetic field acts as a giant spectrometer and enables the determination of the spectral variations of the galactic cosmic rays near Earth and the spectral characteristics of the sporadic solar cosmic rays. In addition, the simultaneous detection of relativistic particles with the entire global network of neutron monitors provides information about the anisotropy of the cosmic ray flux near Earth as the viewing directions of each neutron monitor station at the border of the geomagnetosphere depends on the neutron monitor's location, on the cosmic ray particle's rigidity, and on the direction of incidence above the neutron monitor station.

To deduce the variation of the primary cosmic rays near Earth but outside the geomagnetosphere from neutron monitor measurements, the relationship between the neutron monitor count rate and the primary cosmic ray flux must be known.

Fig. 6.1 18-IGY neutron monitor Jungfraujoch, Switzerland, (*left*) and 6-NM64 neutron monitor Athens, Greece (*right*). The digits 18 respectively 6 give the number of the counter tubes of the corresponding neutron monitor station

[1]Decrease within hours in the galactic cosmic ray intensity near Earth caused by the passage of a coronal mass ejection (CME) and slow recovery within days, named after the American cosmic ray physicist Scott E. Forbush.

When primary cosmic ray particles approach the Earth, they enter first the geomagnetosphere, where the cosmic ray particles are deviated by the Earth's magnetic field (Lorentz force) and then penetrate into the Earth's atmosphere, where the cosmic ray particles make electromagnetic interactions with the atoms and molecules as well as hadronic processes with the nuclei of the atmospheric constituents. A cascade of various secondary particles is produced.

In Sect. 6.2 a short overview of the history of the neutron monitors is given. The transport of cosmic rays in the Earth's magnetic field is described in more detail in Chap. 5. The transport of cosmic ray particles through the Earth's atmosphere is addressed in Sect. 6.3. The neutron monitor, i.e. its structure, layout, the functions of the different parts of the detector, the response of the neutron monitor to primary cosmic rays, and environmental effects on the measurements are described in Sect. 6.4. As a single neutron monitor does not give information about the energy spectrum and the direction of the flux of the primary cosmic rays, a network of neutron monitors at different latitudes and longitudes is needed to retrieve this information. Today this network contains about 50 operating stations. Section 6.5 gives an overview about the worldwide network of neutron monitor stations. The neutron monitor database NMDB initiated in 2008/09 as an European FP7 project is presented in Sect. 6.6.

6.2 History

After the discovery of the cosmic rays in 1912 by Victor Hess, mainly ionisation chambers on ground were used to investigate the variations of the cosmic ray intensities. The basic ideas for the development of neutron monitors as a continuous recorder of the cosmic ray intensity originated from the measurements by Simpson (1948). He found that the latitude dependences of the intensities of the energetic nucleonic component and of the evaporation neutrons from the secondary cosmic rays in the atmosphere are several times larger than those of the ionising component (ionisation chambers) and the hard component (muon counters). In addition, the measurement of the nucleonic component allows to study the time variations of the primary cosmic rays at lower energies than this is possible with ionisation chambers or muon counters. These facts stimulated the development of new detectors that measure the secondary neutrons in the atmosphere.

The neutron monitor designed by Simpson (1955) was adopted as the standard detector during the International Geophysical Year (IGY) 1957/58 and was called IGY neutron monitor. It became evident soon that better statistical accuracy was required, in particular for the study of short-term events as e.g. solar cosmic ray events, so-called GLEs (Ground Level Enhancements or Ground Level Events). In 1959 large sized proportional counter tubes were constructed and produced at the Chalk River Nuclear Laboratories, Ontario, Canada. This led to the design and the construction of the supermonitor or NM64 monitor for the International Quiet Sun Year (IQSY) (Carmichael 1968). The counting rate of the NM64 monitor per unit area of lead producer is about three times that of the IGY neutron monitor. Today mainly NM64 monitors are in operation.

The different neutron monitor stations are mostly operated by research institutions that are located near the stations. For the data exchange in the days of the advent of neutron monitors the operators of the stations sent their hourly data in the form of tables on paper by mail to the World Data Centers (WDCs) (Pyle 2000), later on magnetic tapes and afterward on floppy disks. With the internet the data exchange became much easier.

6.3 Transport of Cosmic Ray Particles in the Earth's Atmosphere

Primary cosmic ray particles that penetrate the atmosphere undergo multiple interactions resulting in showers of secondary particles. If the secondary nucleons (neutrons or protons) reach the ground, they can be detected by neutron monitors.

To deduce the cosmic ray characteristics at the top of the atmosphere from neutron monitor measurements, the transport in the atmosphere, i.e. the interactions of energetic particles with matter, and the detection efficiency of the neutron monitor must be known. The physics of the interactions in the atmosphere when cosmic ray particles enter the atmosphere are today usually simulated with Monte Carlo methods.

The Earth's atmosphere, i.e. the medium in which the interactions take place, is described by a model (Sect. 6.3.1). The essential nuclear interactions of the cosmic ray particles when entering into the atmosphere, which are relevant for ground-based neutron monitor measurements, are addressed in Sect. 6.3.2.

6.3.1 Model of the Earth's Atmosphere

There exist several models that describe the properties (pressure, temperature, density, chemical composition) of the Earth's atmosphere primarily as a function of altitude (US Standard Atmosphere, International Standard Atmosphere, NRLMSISE-00).

Within the Geant4 (Agostinelli et al. 2003) software PLANETOCOSMICS (Desorgher et al. 2006), which is often used to simulate the transport of cosmic ray particles in the atmosphere, it is possible to select between the MSISE-90 model and its upgraded version NRLMSISE-00 (Labitzke et al. 1985; Hedin 1991; Picone et al. 2002). MSIS stands for Mass Spectrometer and Incoherent Scatter Radar, E indicates that the model extends from the ground through the exosphere and the number at the end of the short name is the year of release. NRL stands for the US Naval Research Laboratory. Both models provide temperature, density and concentration profiles vs. altitude from the ground to the exobase (~450–500 km) as function of geographic latitude, longitude, date and time in UT, $F_{10.7}$ index (10.7 cm solar radio flux used as solar UV proxy), $F_{10.7A}$ index (3 month average of $F_{10.7}$) and the geomagnetic index Ap. The dependence of the model on $F_{10.7}$, $F_{10.7A}$, and Ap is neglectable below 80 km. The Earth's atmosphere is divided into superposed homogeneous layers above a solid Earth. The density and composition of the layers are computed according to the altitude or atmospheric depth from the atmospheric model and are constant throughout each layer. The thickness of the layers may be selected by the user. The models may take different geometries: flat or concentric spherical geometry. In the case of the spherical geometry the Earth is modeled by a sphere of 6371.2 km radius and overlying curved layers.

6.3.2 Particle Cascade in the Atmosphere

When primary cosmic ray particles enter the Earth's atmosphere, they make electromagnetic interactions with the atoms and molecules of the atmospheric gases and hadronic interactions with the nuclei of the atmospheric matter. Thereby the cosmic ray particles rapidly loose energy and produce various secondary particles. The mean free path for nuclear interactions of a cosmic ray particle (proton) with nitrogen or oxygen nucleus is ~75 g/cm^2.

For the interpretation of neutron monitor measurements the nucleonic or hadron component in the atmosphere is relevant. The products of the nucleonic interactions are secondary nucleons and pions (π^+, π^-, and π^0). The secondary protons lose their energy mainly by ionisation. When the secondary nucleons have sufficient energy, they continue to multiply in successive generations of nuclear collisions until the energy of the particles drops below the energy that is required for multiple pion production, i.e. about 1 GeV. Secondary protons and ions with energies $\leq$100 MeV no longer undergo hadronic interactions, they are rapidly decelerated to rest by ionisation. On the other hand the neutrons still make nuclear interactions at these energies as well as elastic collisions with nuclei in the atmosphere. Below 10 MeV the neutrons lose their energy continuously by elastic collisions with atmospheric nuclei before they are captured by nucleons at thermal energies.

The neutral pions π^0 have a very short mean lifetime $\tau = 1.78 \times 10^{-16}$ s. The π^0 decays immediately into two γ-rays which initiate an electromagnetic cascade. The charged pions π^+ and π^- decay into muons $\pi^+ \rightarrow \mu^+ + \nu_\mu$ and $\pi^- \rightarrow \mu^- + \overline{\nu}_\mu$ with a mean lifetime of 2.55×10^{-8} s. The muons are slowed down mainly by ionisation. The low energy muons have time to decay ($\tau = 2.2 \times 10^{-6}$ s) before they reach the ground. The reactions are $\mu^+ \rightarrow e^+ + \nu_e + \overline{\nu}_\mu$ and $\mu^- \rightarrow e^- + \overline{\nu}_e + \nu_\mu$. However, many of the muons are produced with very high energies in the uppermost layers of the atmosphere and as these muons loose only little energy they have a large path length and survive (time dilatation according to the theory of relativity) and reach the surface of the Earth. The muons deep in the atmosphere and at sea level are the most dominant component of the secondary cosmic rays and are therefore the dominant source of ionisation in this altitude range.

The primary particles at the top of the atmosphere must have an energy of roughly 500 MeV per nucleon to produce a cascade of secondary nucleons that can reach the ground at sea level. In high latitude regions, where the shielding effect of the Earth's magnetic field for incident cosmic rays is low, the lower threshold of the neutron monitor response is controlled therefore by the atmospheric mass which is $\sim$1030 g cm^{-2} at sea level. For high latitude neutron monitors at sea level this atmospheric cutoff for primary cosmic rays is $\sim$450 MeV/nucleon. In regions like Central Antarctica, at an elevation of $\sim$3 km above sea level (asl), the reduced atmospheric mass lowers the threshold to $\sim$300 MeV/nucleon (Mishev et al. 2014).

6.4 Neutron Monitor Detector

Neutron monitors cover the energy range of primary cosmic ray particles from $\sim$0.5 to $\gtrsim$100 GeV per nucleon. This energy range includes the solar modulation of the galactic cosmic rays, Forbush decreases, sporadic GLEs or relativistic SEP (Solar Energetic Particle) events, and geomagnetic effects. The longterm stability of neutron monitors is generally excellent so that the cosmic ray effects of the 11-year solar activity cycle can be investigated over several solar cycles. The longest time series of single neutron monitor stations are available over a period of $\sim$60 years, i.e. during a time range over more than five solar activity cycles. The comparison of the measurements of different neutron monitor stations have however shown that some neutron monitors may show a degrading of the detector efficiency (Bieber et al. 2007).

As the name suggests, neutron monitors record predominantly the secondary neutrons from the atmospheric cascades. The contribution to the total neutron monitor counting rate of an NM64 are neutrons $\sim$85%, protons $\sim$7%, μs $\sim$6%, πs $\sim$1% (Hatton 1971).

The functionality, construction and other properties of a neutron monitor are described in Sect. 6.4.1. The response of the neutron monitor to primary cosmic ray particles at the top of the Earth's atmosphere is explained in Sect. 6.4.2. The influence of the atmospheric effects on the neutron monitor measurements are addressed in Sect. 6.4.3.

6.4.1 Components of a Neutron Monitor

Both neutron monitor types IGY (Simpson 1955) and NM64 (Carmichael 1964) employ the same measurement strategy, i.e. the difference in the way high and low energy neutrons interact with different nuclei. As a particle with no electric charge, the neutron makes only interactions with nuclei and can therefore penetrate large layers of material without interactions because of the small range of the strong nuclear force. Energetic neutrons can make three different kinds of interactions with nuclei: elastic and inelastic collisions as well as nuclear reactions. After a hadronic interaction of an energetic neutron with a nucleus, the excited target nucleus emits so-called evaporation neutrons. The production of these evaporation neutrons is proportional to $A^{2/3}$ according to the nuclear physics theory, where A is the atomic weight of the target nucleus. In a material containing nuclei with low atomic mass, the neutrons are effectively slowed down (moderated) in elastic collisions.

These facts led Simpson to the neutron monitor detector concept: production of fast neutrons in a target with high atomic weight, braking of the fast neutrons in a hydrogenous material, and finally detection of the thermic neutrons indirectly by ionising particles that are produced in a neutron induced nuclear reaction.

Figure 6.2 shows a schematic diagram of the NM64 neutron monitor. A standard NM64 neutron monitor with six counter tubes (6-NM64) has the following dimensions: width: $\sim$315 cm, depth: $\sim$220 cm, and height: $\sim$50 cm. The lead producer with $\sim$9650 kg is by weight the major component of an 6-NM64 monitor.

The different components of a neutron monitor detector are described in the following. The given specifications are valid for the NM64 neutron monitor (Carmichael 1964):

Reflector The whole assembly of the detector is enclosed by polyethylene (proton-rich material) of an average thickness of 7.5 cm. The task of the reflector is to reflect and to moderate the evaporation neutrons that are produced in the lead producer. The polyethylene (moderator material) contains a significant fraction of hydrogen, i.e. the energy loss per elastic collision of a neutron with the moderator material is maximal as the mass of the projectile and of the hydrogen nuclei in the target material are almost equal (conservation of momentum and energy). The neutron elastic interaction pathlength with hydrogen in polyethylene is roughly 1 cm for

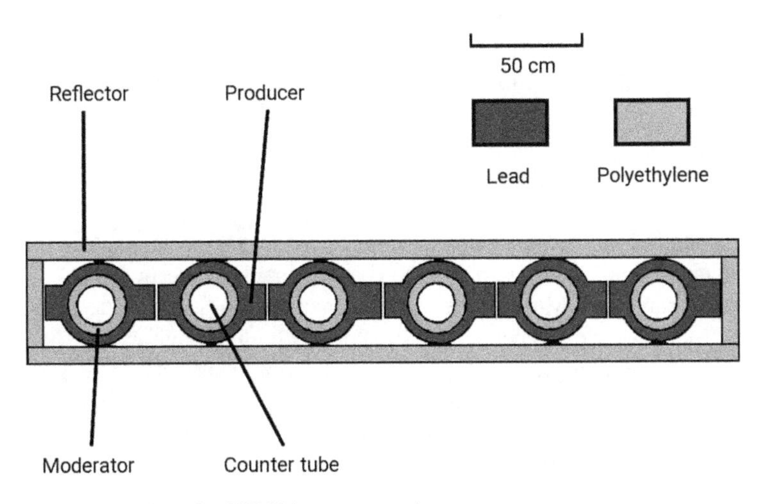

Fig. 6.2 Schematic view of a 6-NM64 neutron monitor

neutrons with energies $\leq$1 MeV and in each collision the incident neutron reduces its kinetic energy on average by a factor of 2, i.e. the evaporation neutrons are very effectively slowed down in the reflector.

In addition, this neutron monitor component has the function to reflect and to absorb the low energy neutrons that are produced by high energy nucleons in interactions with the ambient material of the neutron monitor, e.g. detector housing.

In contrast, this reflector is largely transparent to the energetic neutrons that are produced in the cosmic ray induced cascade in the atmosphere, i.e. these energetic neutrons can easily reach and enter the lead producer.

Producer The core of the neutron monitor consists of a lead producer, a target with high atomic mass (A), to produce secondary neutrons. The average depth of 13.8 cm (156 g/cm^2) corresponds to about 75% of the inelastic mean-free path of nucleons in lead. Thus ~50% of the nucleons, that cross the reflector and enter the lead, make at least one interaction in the producer. The production rate of neutrons per inelastic nucleon-nucleus interaction is roughly proportional to $A^{0.7}$ in the energy range 100–700 MeV of the interacting nucleon and slowly decreases with increasing energy (Shen 1968). The average number of produced neutrons (multiplicity) depends weakly on the incident nucleon energy, i.e. a neutron monitor can be used as an energy spectrometer by measuring the multiplicity only limited. On average about 15 evaporation neutrons with mean energy ~2.5 MeV are produced per nuclear reaction (Hatton 1971). These neutrons amplify the cosmic ray signal and can not easily escape the reflector. The lead producer is interspersed with a moderator and the BF_3 proportional counter tubes.

Moderator Each counter tube is surrounded by a polyethylene tube with a thickness of 2 cm acting as a moderator for the evaporation neutrons that are generated in the lead producer.

Proportional Counter The proportional counter tubes are filled with BF_3 (boron trifluoride) as counter gas. The BF_3 has been 90% enriched with the ^{10}B isotope. When the very slow neutrons (thermal neutrons, $E = \frac{3}{2}\,k\,T = \sim 0.04$ eV) encounter a $^{10}B_5$ nucleus in the proportional counter, the following favoured reaction may take place:

$$^{10}B_5 \quad + \quad n \quad \rightarrow \quad {}^{7}Li_3 \quad + \quad \alpha \tag{6.1}$$

The cross-section for this reaction is inversely proportional to the neutron speed and has a value of about $3.0 \cdot 10^{-25}$m^2 or 3000 barns at neutron energy 0.04 eV and only roughly 0.2 barns at 1 MeV (Clem and Dorman 2000). The produced α-particle and the Li-nucleus are accelerated by the applied high voltage within the counter tube, ionize the counter gas and the produced electrons cause an electric signal. The electric signal is amplified, discriminated and counted by a counter electronic. The detection probability of the evaporation neutrons is ~5.7% (Hatton 1971).

Later proportional counter tubes were filled with ^{3}He as an alternative to the standard BF_3 counters, as BF_3 is highly toxic. The He counters require a higher pressure to have an efficiency close to the BF_3 counters and they have a much higher temperature sensitivity and therefore require better environmental temperature stability. Currently, new neutron monitor counter tubes use again BF_3 because of the very high price for ^{3}He.

As an incident neutron or proton into the neutron monitor may produce more than one evaporation neutron in the lead producer, it can be expected that a group of count signals is observed (multiplicity). However, the efficiency for detecting these evaporation neutrons is low. Therefore, the average detected multiplicity is typically not much larger than one. Due to the multiplicity effect in the neutron monitor the count impulses are not equally distributed. Therefore the variance of the counting rate $\sigma(N)$ is not Poisson distributed. One has $\sigma(N) > \sqrt{N}$. The variance of the counting rate $\sigma(N)$ is given by:

$$\sigma(N) = k \cdot \sqrt{N} \tag{6.2}$$

where k is around 1.5 for an NM64 monitor (Hatton 1971).

The average count rate of an 6-NM64 monitor at high latitude and at sea level is ~4200 counts per minute. Due to the multiplicity effect in the neutron monitor the relative random error for 1-min values is therefore ~2.5%. The count rate of an equatorial sea-level 6-NM64 monitor is ~0.7 times the count rate of an identical neutron monitor at high latitudes. Neutron monitors at high altitudes have higher counting rates because of the smaller atmospheric attenuation. The count rate of a neutron monitor at high latitudes and at an altitude of ~3000 m asl is about a factor of ten higher than at sea level.

6.4.2 Neutron Monitor Yield Function

The transport of cosmic ray particles through the Earth's atmosphere and the detection of the nucleonic component of the secondary cosmic rays by the neutron monitors are combined in the so-called neutron monitor yield function. The neutron monitor yield function can therefore directly be used to determine the cosmic ray flux at the top of the Earth's atmosphere from the measurements of the worldwide neutron monitors. Essentially, two methods are used to determine the neutron monitor response function:

- parameterisation of latitude survey observations (neutron monitor measurements e.g. on a ship cruise along a large range of geomagnetic latitudes)
- Monte Carlo simulations of the cosmic ray transport through the Earth's atmosphere and of the detection efficiency for the different secondary particles in the neutron monitor

The most commonly used response function based on latitude surveys is the Dorman function (Dorman and Yanke 1981). The Dorman function represents most latitude surveys fairly well, however the trend of the response function at low rigidities can not be retrieved from latitude surveys. Belov and Struminsky (1987) made modifications to the Dorman function for rigidities <2.78 GV.

Different authors have determined the neutron monitor yield function with Monte Carlo simulations. In 1982 Debrunner et al. calculated the specific yield function for sea level neutron monitors. Clem and Dorman (2000) applied the FLUKA Monte-Carlo package (Fassò et al. 1993) for the simulations. In recent years different groups (Flückiger et al. 2008; Matthiä 2009; Mishev et al. 2013) used the Geant4 software package (Agostinelli et al. 2003) to compute the neutron monitor yield function. For the transport in the atmosphere the Geant4 software suite PLANETOCOSMICS (Desorgher et al. 2006) has been mostly used. Some of the determined yield functions are valid only for neutron monitors at sea level. This requires pressure corrections of the neutron monitor count rates to sea level. Mainly for high altitude neutron monitor stations these corrections may be inaccurate as the parameter for the barometric corrections depends on the cosmic ray spectrum, see Sect. 6.4.3. The yield function by Flückiger et al. (2008) is valid for different altitudes, i.e. the atmospheric depth of the neutron monitor station is a parameter of this yield function.

The relation between the counting rate N_x of the neutron monitor station x and the differential fluxes of the different components k of the primary cosmic rays at the border of the geomagnetic field, ψ_k, can be described by the following formula:

$$N_x(h, \Lambda, \chi, t) = \sum_k \int_0^{2\pi} \int_0^{\pi/2} \int_0^{\infty} \alpha(R, \theta, \phi, \Lambda, \chi, t) \cdot \psi_k(R, \theta, \phi, \Lambda, \chi, t) \cdot \sin\theta \cdot S_x^k(h, R, \theta) \cdot A_x \cdot \cos\theta \cdot dR\, d\theta\, d\phi \tag{6.3}$$

where

- α filter function which is 1 for allowed cosmic ray particle trajectories and 0 for forbidden trajectories (Chap. 5).
- R rigidity of the primary cosmic ray particle. $R = \frac{p \cdot c}{Z \cdot e}$ where p is the momentum of the particle, Ze is its charge and c is the speed of light.
- θ, ϕ zenith and azimuth angle of the primary cosmic rays at the top of the atmosphere.
- Λ, χ geographic latitude and longitude of the neutron monitor station location.
- t date and time in UT.
- S_x^k yield function. It gives the number of count events produced by a primary particle of type k in the neutron monitor station x. The yield function depends on the rigidity, on the zenith angle, θ, of the incident particle, and on the atmospheric depth, h, of the neutron monitor station.
- h atmospheric depth of the neutron monitor station.
- A_x area of the neutron monitor station, x.

6.4.3 Atmospheric Effects

Because of the interactions of the primary and secondary cosmic ray particles with the matter of the Earth's atmosphere, the neutron monitor count rate depends also on meteorological conditions (Carmichael et al. 1968). With constant cosmic ray intensity at the top of the Earth's atmosphere the counting rate of a neutron monitor depends mainly on the atmospheric mass above the detector and in much lower degree on the temperature profile and on the water content in the atmosphere. Because of the relative small effects and because of the large complexity to determine the temperature profile and the water content in the atmosphere, only the change in the atmospheric mass is considered for neutron monitor measurements. In practice, the barometric pressure is used as a proxy for the air mass to correct the neutron monitor data to a constant atmospheric depth.

The dependence of the neutron monitor count rate upon atmospheric pressure is usually described by an exponential function:

$$N(p(t)) = N(\overline{p}) \cdot \exp\left(\frac{\overline{p} - p(t)}{\lambda}\right) \tag{6.4}$$

where

$N(p(t))$ measured count rate at atmospheric pressure p and at time t.
$N(\overline{p})$ neutron monitor count rate at some standard pressure $\overline{p}$.
λ attenuation length of the nucleonic component of the cosmic radiation in the Earth's atmosphere.

The attenuation length λ depends on the altitude, the geomagnetic latitude and on the primary cosmic ray spectrum. The attenuation length is larger for harder primary rigidity spectra and vice versa.

During a solar cosmic ray event measured by neutron monitors, the cosmic ray near Earth includes a galactic and a solar component. The solar particles show a softer rigidity spectrum compared to the spectrum of the galactic cosmic rays. The attenuation length for galactic cosmic rays λ_g is about 140 g/cm^2, whereas the value for the solar cosmic rays λ_s is typically around 100 g/cm^2. The two-attenuation length method by McCracken (1962) considers this fact.

The pressure corrected neutron monitor counting rate to the selected standard pressure $\overline{p}$ during a solar cosmic ray event can be written as follows by assuming that the galactic cosmic ray intensity during the solar cosmic ray event does not change:

$$\begin{aligned} N(t,\overline{p}) &= N_0(p_0) \cdot \exp\left(\frac{p_0 - \overline{p}}{\lambda_g}\right) + \\ &\left[N\left(t, p(t)\right) - N_0(p_0) \cdot \exp\left(\frac{p_0 - p(t)}{\lambda_g}\right)\right] \cdot \exp\left(\frac{p(t) - \overline{p}}{\lambda_s}\right) \end{aligned} \tag{6.5}$$

where

$N(t, \overline{p})$	pressure corrected count rate to the standard pressure $\overline{p}$ of the neutron monitor station during a solar cosmic ray event.
$N_0(p_0)$	average measured count rate with an average atmospheric pressure p_0 during the reference time interval (typically the full hour before the onset of the solar cosmic ray event).
λ_g	attenuation length for the nucleonic component of the galactic cosmic rays in the Earth's atmosphere.
$N(t, p(t))$	measured count rate at time t and atmospheric pressure $p(t)$.
λ_s	attenuation length for the nucleonic component of the solar cosmic rays in the Earth's atmosphere.

The first summand of equation (6.5) is the contribution to the count rate by the galactic cosmic ray corrected to the standard pressure $\overline{p}$. The second summand is the pressure corrected count rate caused by the solar cosmic rays. The expression between square brackets is the measured part to the counting rate caused by the solar cosmic ray at time t. For the determination of this portion the measured counting rate during the reference time interval $N_0(p_0)$ has to be corrected to the current atmospheric pressure $p(t)$ and has to be subtracted from the measured count rate $N(t, p(t))$ during the GLE.

The barometric pressure coefficient for galactic cosmic rays $\alpha = \lambda^{-1}$ for a neutron monitor has a value in the order of 1%/mmHg (or $0.0072\,\mathrm{mbar}^{-1}$), i.e. the change in the air mass above a neutron monitor station has a large effect upon the count rate. Therefore, the barometric pressure at a neutron monitor station must be determined very accurately, as an error in the pressure measurement of 1 mmHg causes a change in the count rate of $\sim$1%. As the spectrum of the galactic cosmic ray changes during the 11-year solar activity cycle, the barometric coefficient α shows a variation as well and should therefore be determined periodically.

Neutron monitor stations at exposed locations (e.g. high altitudes) may considerably be affected by environmental effects. These neutron monitor stations are heavily exposed to high wind speeds and gusty winds which may strongly affect the atmospheric pressure measurements. Consequently, the correction of the neutron monitor count rates for the effects of changes in the air mass above the detector using raw barometer data may lead to erroneous results (Bütikofer and Flückiger 1999). In addition, there are other environmental effects on the neutron monitor counting rate, e.g. the accumulation of snow on the roof and around the detector housing. This effect must usually not be taken into account during a solar cosmic ray event. However, when investigating long time data series, the use of neutron monitor stations, where possible changes of snow accumulations may occur on the roof and around of the detector housing, must be considered with care. These effects must be considered especially for the NM64 type as the thickness of the reflector is only 7.5 cm compared to 28 cm in the IGY neutron monitor. In the NM64 monitor the evaporation neutrons, that are produced in the surrounding material of the neutron monitor, contribute to the counting rate with ∼5% (Hatton 1971) and changes of this matter in the immediate environment therefore affect the counting rate.

6.5 Worldwide Network of Neutron Monitor Stations as a Giant Spectrometer

The ‘Simpson’ neutron monitor (Simpson 1955) was the standard cosmic ray detector for the International Geophysical Year (IGY) 1957/58, and it was called the ‘IGY’ neutron monitor. During the years 1957–1959 a worldwide network of 51 monitors was established. After the International Geophysical Year 1957/58 some of the IGY neutron monitors stopped, however most stations continued operating. With the launch of the NM64 or ‘supermonitor’ in the 1960s by Carmichael (1968) many of the IGY neutron monitors were replaced by the new detector type. During this transition time most principal investigators operated both neutron monitor types in parallel for some months to determine a normalization factor for long term studies. Today the majority of the worldwide network comprises NM64 monitors, however there are still a few IGY neutron monitors in operation. Figure 6.3 shows a world map with the locations of the neutron monitor stations that have been in operation in 2017 or only recently been closed. In 2017 about 50 neutron monitor stations have been in operation.

The Earth’s magnetic field establishes the worldwide network of neutron monitors to a huge spectrometer. The rigidity range of this spectrometer is determined by the atmospheric cutoff at the lower rigidity border and by the highest magnetic cutoff rigidity at the other end. Although the magnetic cutoff rigidity near the geomagnetic

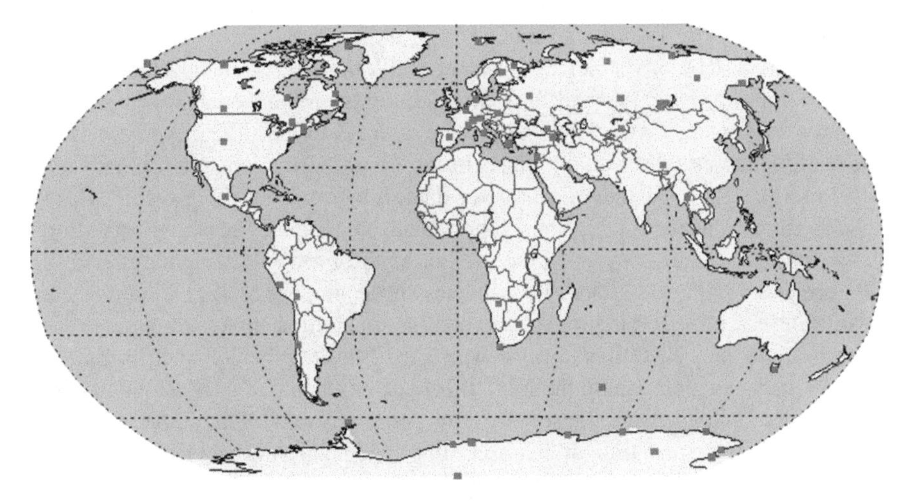

Fig. 6.3 World map with the locations of neutron monitors which have been in operation in 2017 or only recently been closed

poles is 0 GV as the magnetic field lines enter vertically into the Earth, the primary cosmic ray particles penetrating the top of the atmosphere must have a minimal energy that the secondary nucleons can reach the ground. This atmospheric cutoff energy for a sea level detector is $\sim$450 MeV, i.e. a rigidity of $\sim$1 GV for protons. The maximum vertical magnetic cutoff rigidity is $\sim$15 GV. The measurements of the worldwide network of neutron monitors enable therefore to determine spectral variations of the galactic cosmic rays near Earth and the spectral characteristics of GLEs in the energy range from $\sim$500 MeV to $\sim$15 GeV.

When working with neutron monitor data it is important to realise that the neutron monitor stations of the worldwide network are operated by different institutes, i.e. the measurements (e.g. obvious outliers in the count rate, data gaps etc.) are handled differently. The neutron monitor stations are located at very different locations (sea level, high altitude, polar regions), therefore e.g. the stability of the temperature inside the detector housing, of the humidity as well as the behaviour of the electronic devices may differ and cause different qualities of the measurements. In earlier days there were also problems with the accuracy of the used clocks.

6.6 Neutron Monitor Database: NMDB

In the early days of neutron monitors, the cosmic ray scientists exchanged their data with data tables and books (mostly on a monthly/half-year basis and only with a resolution of one hour). Later the operators of neutron monitor stations wrote the measured data on storage media like magnetic tapes or floppy disks and sent these media to the World Data Centers (WDC) in the USA, USSR, and Japan (Pyle 2000). For neutron monitor data analysis the scientists ordered the data either directly from the PIs or from the WDCs and received the data on magnetic tape and later on compact discs. The advent of the internet made the data exchange much easier. The different groups published their measurements on their own webpage. More and more the cosmic ray scientists have the demand to have the measurements of the worldwide network available in real-time. There were also some initiatives to develop a data base for neutron monitor data, however only the project "NMDB – Real-Time database for high resolution Neutron Monitor measurements" (http:\www.nmdb.eu) funded by the Commission of the European Communities as an FP7 project in the years 2008/09 was successful. A number of 12 institutions were involved in the project. First only the data of the institutions involved in the project were brought into NMDB. However, since then the number of neutron monitor stations that send their data to the NMDB has increased and in 2017 a total of about 40 neutron monitor stations transmit their data regularly to NMDB, about 30 neutron monitor stations in real-time or near real-time. The neutron monitor measurements are stored in NMDB as 1-min and hourly data.

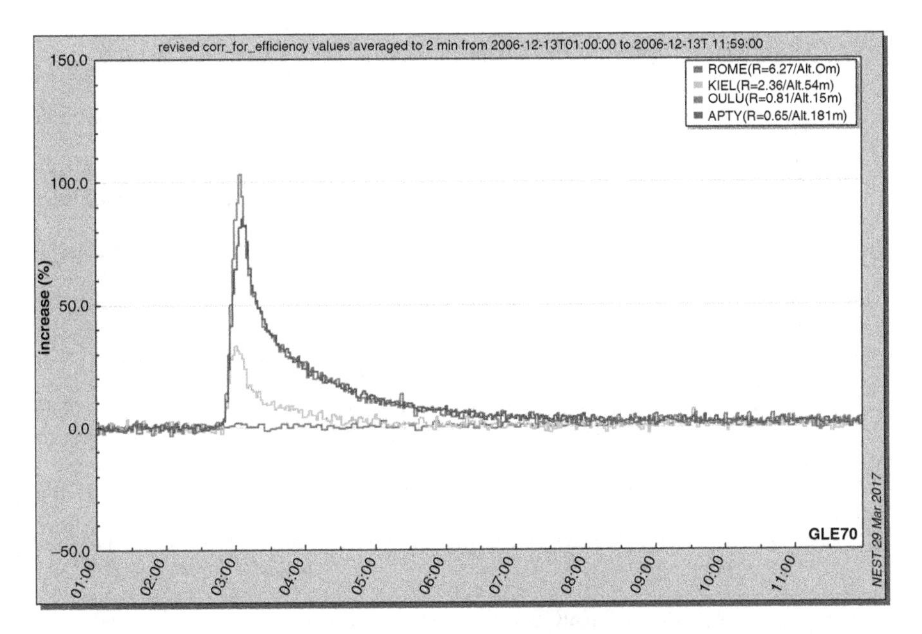

Fig. 6.4 Relative increase of pressure corrected 2-min data of the neutron monitor stations Oulu, Apatity, Kiel and Rome during GLE#70 (13 December 2006) plotted with NEST

In addition to the database there are different applications available from the NMDB webpage www.nmdb.eu (Mavromichalaki et al. 2011). NMDB provides e.g. the application NEST (http://www.nmdb.eu/nest/) to generate plots of the count rate of selected neutron monitor stations with different time resolutions. It is also possible to plot neutron monitor data together with the sunspot number (smoothed or monthly), geomagnetic *Kp*-index (3-hourly), or GOES data (channels >10 MeV, >50 MeV, >100 MeV). The plots can be modified by different style adjustments. The generation of plots during GLEs and Forbush decreases from the past can be selected with one click. Figure 6.4 shows as an example the measured relative increase in the count rate of a selection of neutron monitors during GLE#70 on 13 December 2006 as plotted with the NMDB NEST application. In addition to the graphic output it is also possible to extract ASCII data from NMDB with NEST. Other NMDB applications are e.g. GLE alarm systems or GLE characteristics determination.

In addition to NMDB the Cosmic Ray Station of the University of Oulu reactivated and operates the GLE database http://gle.oulu.fi/ where the neutron monitor data of the worldwide network during GLEs are stored and are made available for plotting and for downloading.

References

Agostinelli, S., et al.: GEANT4a simulation toolkit. Nucl. Instrum. Methods Phys. Res. A **506**, 250–303 (2003)

Belov, A.V., Struminsky, A.B.: Neutron monitor sensitivity to primary protons below 3 GeV derived from data of ground level events. In: 25th International Cosmic Ray Conference, Durban, vol. 1, p. 201 (1987)

Bieber, J.W., Clem, J., Desilets, D., Evenson, P., Lal, D., Lopate, C., Pyle, R.: Long-term decline of South Pole neutron rates. J. Geophys. Res. (Space Phys.) **112**, A12102 (2007)

Bütikofer, R., Flückiger, E.O.: Pressure Correction of GLE measurements in turbulent winds. Int. Cosmic Ray Conf. **6**, 395 (1999)

Carmichael, H.: Cosmic Rays, IQSY Instruction Manual No. 7. IQSY Secretariat, London (1964)

Carmichael, H.: Annals of the IQSY. In: Minnis, C.M. (ed.) Geophysical Measurements: Techniques, Observational Schedules and Treatments of Data, vol. 1. MIT Press, Cambridge (1968)

Carmichael, H., Bercovitch, M., Shea, M.A., Magidin, M., Peterson, R.W.: Attenuation of neutron monitor radiation in the atmosphere. Can. J. Phys. Suppl. **46**, 1006 (1968)

Clem, J.M., Dorman, L.I.: Neutron monitor response functions. Space Sci. Rev. **93**, 335–359 (2000)

Debrunner, H., Lockwood, J.A., Flückiger, E.: Specific yield function S(P) for a neutron monitor at sea level, paper presented. In: 8th European Cosmic Ray Symposium, Rome (1982)

Desorgher, L., Flückiger, E.O., Gurtner, M.: The PLANETOCOSMICS Geant4 application. In: 36th COSPAR Scientific Assembly, vol. 36 of COSPAR Meeting, p. 2361 (2006)

Dorman, L.I., Yanke, V.G.: The coupling functions of NM-64 neutron supermonitor. Int. Cosmic Ray Conf. **4**, 326 (1981)

Fassò, A., Ferrari, A., Ranft, J., Sala, P.R., Stevenson, G.R., Zazula, J.M.: A comparison of FLUKA simulations with measurements of fluence and dose in calorimeter structures. Nucl. Instrum. Methods Phys. Res. A **332**, 459–468 (1993)

Flückiger, E.O., Moser, M.R., Pirard, B., Bütikofer, R., Desorgher, L.: A parameterized neutron monitor yield function for space weather applications. Int. Cosmic Ray Conf. **1**, 289–292 (2008)

Hatton, C.J.: The Neutron Monitor. American Elsevier Publishing Company, New York (1971)

Hedin, A.E.: Extension of the MSIS thermosphere model into the middle and lower atmosphere. J. Geophys. Res. **96**, 1159–1172 (1991)

Labitzke, K., Barnett, J.J., Edwards, B.: Handbook MAP, vol. 16 (1985)

Matthiä, D.: The radiation environment in the lower atmosphere: a numerical approach. Ph.D. thesis, Christian-Albrechts-Universität zu Kiel (2009)

Mavromichalaki, H., et al.: Applications and usage of the real-time neutron monitor database. Adv. Space Res. **47**, 2210–2222 (2011)

McCracken, K.G.: The cosmic-ray flare effect, 1, some new methods of analysis. J. Geophys. Res. **67**, 423–434 (1962)

Mishev, A.L., Usoskin, I.G., Kovaltsov, G.A.: Neutron monitor yield function: new improved computations. J. Geophys. Res. (Space Phys.) **118**, 2783–2788 (2013)

Mishev, A.L., Kocharov, L.G., Usoskin, I.G.: Analysis of the ground level enhancement on 17 May 2012 using data from the global neutron monitor network. J. Geophys. Res. (Space Phys.) **119**, 670–679 (2014)

Picone, J.M., Hedin, A.E., Drob, D.P., Aikin, A.C.: NRLMSISE-00 empirical model of the atmosphere: statistical comparisons and scientific issues. J. Geophys. Res. (Space Phys.) **107**, 1468 (2002)

Pyle, R.: Public access to neutron monitor datasets. Space Sci. Rev. **93**, 381–400 (2000)

Simpson, J.A.: The latitude dependence of neutron densities in the atmosphere as a function of altitude. Phys. Rev. **73**, 1389–1391 (1948)

Simpson, J.A.: Cosmic radiation neutron intensity monitor. Ann. Int. Geophys. Year **4**, 351–373 (1957)

Shen, M.: Neutron production in lead and energy response of neutron monitor. Suppl. Nuovo Cimento **6**, 1177 (1968)

Chapter 7
HESPERIA Forecasting Tools: Real-Time and Post-Event

Marlon Núñez, Karl-Ludwig Klein, Bernd Heber, Olga E. Malandraki, Pietro Zucca, Johannes Labrens, Pedro Reyes-Santiago, Patrick Kuehl, and Evgenios Pavlos

Abstract Within the HESPERIA Horizon 2020 project, two novel real-time tools to predict Solar Energetic Particle (SEP) events were developed. The HESPERIA UMASEP-500 tool makes real-time predictions using a lag-correlation between the soft X-ray (SXR) flux and high-energy differential proton fluxes of the GOES satellite network. We found that the use of proton data alone allowed this tool to make predictions before any Neutron Monitor (NM) station's alert. The performance of this tool for predicting Ground Level Enhancement (GLE) events for the period 2000–2016 may be summarized as follows: the probability of detection (POD) was 53.8%, the false alarm ratio (FAR) was 30%, and the average warning time (AWT) to the first NM station's alert was 8 min. The developed HESPERIA REleASE tool makes real-time predictions of the proton flux-time profiles of 30–50 MeV protons at L1 and is based on electron intensity measurements of energies from 0.25 to 1 MeV and their intensity changes. The performance was tested by using all historic ACE/EPAM and SOHO/EPHIN data from 2009 until 2016 and has shown that the forecast tools have a low FAR ($\sim$30%) and a high POD (63%). Furthermore, two methods using historical data were explored for predicting SEP events and

M. Núñez (✉) • P. Reyes-Santiago
Universidad de Málaga, Málaga, Spain
e-mail: mnunez@uma.es

K.-L. Klein • P. Zucca
Observatoire de Paris, Meudon, France
e-mail: ludwig.klein@obspm.fr

B. Heber • J. Labrens • P. Kuehl
Christian-Albrechts – University of Kiel, Kiel, Germany
e-mail: heber@physik.uni-kiel.de

O.E. Malandraki • E. Pavlos
National Observatory of Athens, IAASARS, Athens, Greece
e-mail: omaland@astro.noa.gr

O.E. Malandraki, N.B. Crosby (eds.), *Solar Particle Radiation Storms Forecasting and Analysis, The HESPERIA HORIZON 2020 Project and Beyond*, Astrophysics and Space Science Library 444, DOI 10.1007/978-3-319-60051-2_7

compared. The UMASEP-10mw tool was developed for predicting >10 MeV SEP events using microwave data. The time derivative of the soft X-rays (SXR) was replaced by the microwave flux density. It was found that the use of SXRs and microwave data produced the same POD ($\sim$78%) with the most notable difference being that the use of microwave data does not yield any false alarm. Furthermore, a study was carried out on the possibility for the microwave emissions to be used to predict the spectral hardness of the SEP event and important results were deduced.

7.1 Introduction

Forecasting solar energetic particle (SEP) events is of potential interest for spacecraft and launching operations, and for the assessment of radio wave propagation conditions in the polar ionosphere of the Earth. It will be mandatory for human spaceflight beyond low-Earth orbit, especially outside the Earth's magnetosphere. Besides predicting SEP events in general, the prediction of particularly energetic SEPs is a second aim of forecasting, because they penetrate deeper into the terrestrial atmosphere and contribute to the radiation dose aboard aircraft.

Operational real-time SEP forecasts are currently supported by empirical models which rely on observations of associated solar phenomena, including electromagnetic signatures of SEP acceleration/escape near the Sun and observations at the near-Earth environment (L1 or 1 AU) of energetic particles. In this chapter the two novel real-time SEP forecasting tools developed and operating within the HESPERIA project are presented, based on the University of MAlaga Solar particle Event Predictor (UMASEP) (Núñez 2011, 2015) and Relativistic Electron Alert System for Exploration (REleASE) schemes (Posner 2007).

The developed and operational HESPERIA UMASEP-500 tool makes real-time predictions of the occurrence of Ground Level Enhancement (GLE) events, from the analysis of soft X-ray (SXR) and differential proton flux measured by the Geostationary Operational Environmental Satellites (GOES) satellite network. Using near-relativistic as well as relativistic electrons as precursors for the arrival of energetic protons, the developed HESPERIA REleASE tools make real-time predictions of the proton flux-time profiles of 30–50 MeV protons at L1. Furthermore, two methods using historical data explored under the HESPERIA project for predicting SEP events are presented and compared. We have tested if the UMASEP scheme can be improved using microwave observations and also studied if the microwave emissions can be used to predict the spectral hardness of the SEP event and important results are deduced.

The first two sections of the chapter are dedicated to the investigations of whether historical microwave emissions can be used in the forecasting of SEP events (Sect. 7.2 and 7.3), whereas the two following sections describe the real-time HESPERIA SEP forecasting tools, using the REleASE and UMASEP proven concepts (Sect. 7.4 and 7.5). Concluding remarks as well as future possibilities are given in Sect. 7.6.

7.2 Predicting SEP Event Onsets from Historical Microwave Data by Using the UMASEP Scheme

Within the HESPERIA project we tested whether microwave emission could be used in the forecasting of SEP events. This is not possible in real time, because no real-time microwave data are presently provided. However, we attempted a proof-of-concept by using historical SEP events.

The UMASEP forecasting scheme (Núñez 2011, 2015) uses the positive time derivative of the observed SXR flux as an indicator of energy release at the Sun. The SXR burst shows the heating of the corona during a flare. The UMASEP scheme considers that a common positive derivative with the particle flux near Earth, with a suitable time delay, indicates a magnetic connection between the Earth and a site of particle acceleration near the Sun. It is well known (Neupert 1968; Dennis and Zarro 1993; Holman et al. 2011) that hard X-ray (HXR) or microwave bursts, produced by non-thermal electrons in the solar atmosphere through bremsstrahlung and gyrosynchrotron emission (see Chap. 2), have time profiles that mimic the time derivative of the SXR. The reason is a common time evolution of the energy release that goes to the electron acceleration on the one hand and to the heating of the plasma during the related flare on the other. While non-thermal electrons lose their energy rapidly through interactions with the solar atmosphere, the heated coronal plasma cools on much slower time scales, and its time evolution is therefore the integral over the distinct episodes of energy release traced by the non-thermal signatures. So as long as the microwave emission is due to gyrosynchrotron radiation of non-thermal electrons, its time profile can be considered as being close to the time derivative of the SXR profile in the impulsive flare phase.

Patrol observations of the whole Sun at microwaves can be conducted with ground-based antennas. The US Air Force operates the Radio Solar Telescope Network (RSTN) consisting of four stations around the world. These stations observe independently from each other, but with identical equipment at selected frequencies.

The most interesting frequencies for our purpose are 4.995 (henceforth referred to as 5 GHz), 8.8 and 15.4 GHz. The data are publicly available after about 1 year via the National Geophysical Data Center (NGDC).[1] Data from the Nobeyama Radio Polarimeters[2] (NoRP), (Torii et al. 1979; Nakajima et al. 1985), operated by the National Astronomical Observatory of Japan, were used for checking purposes and to replace RSTN when necessary.

In order to test to which extent microwave data can support the UMASEP scheme, we constructed a continuous time series of RSTN observations during a 13-month long period from December 2011 to December 2012. The observations

[1] http://www.ngdc.noaa.gov/stp/space-weather/solar-data/solar-features/solar-radio/rstn-1-second/
[2] http://solar.nro.nao.ac.jp/norp/html/event/

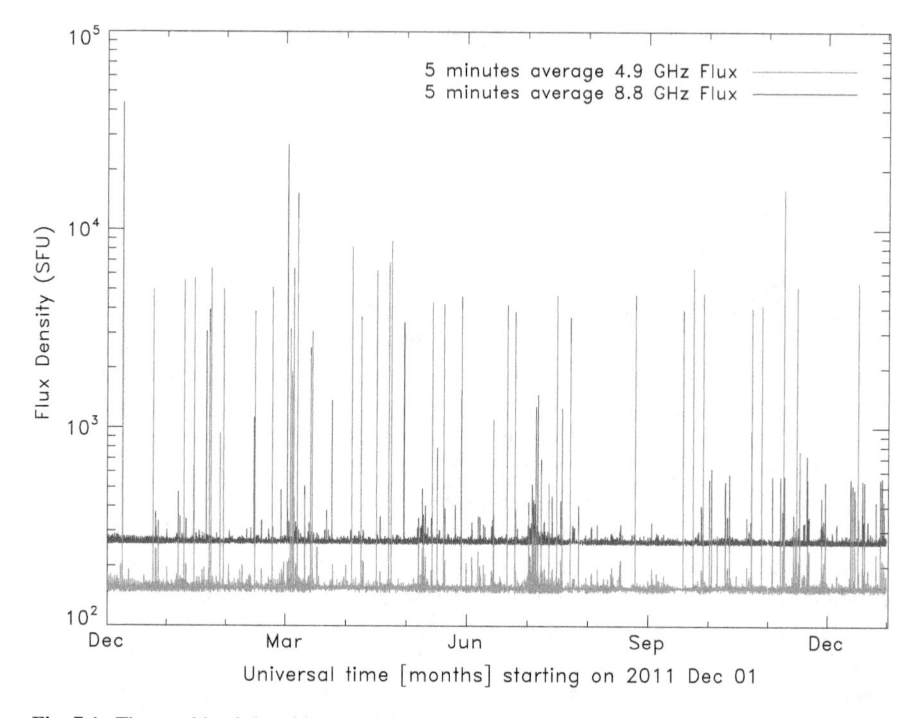

Fig. 7.1 The combined time history of the microwave flux density at two frequencies during the 13 months from 2011 Dec 01 to 2012 Dec 31, constructed from observations of the four RSTN stations

of the individual RSTN stations have a number of problems that needed a careful consideration. We applied a series of standard treatments to remove spikes, data gaps, baseline drifts due to wrong antenna pointing. Then the corrected daily records of the individual stations were combined into the 13-month long time series. A uniform average background was added at each frequency, and smaller flux densities were set to the background value. 5-min integration further smoothes out short-term irregularities that remain after the data cleaning procedure.

Figure 7.1 shows the resulting flux density calculated for the 13 months interval from December 2011 to December 2012. At both frequencies numerous bursts are seen. During this period, nine SEP events were considered as well-connected events and four were considered as poorly connected. An SEP event in the sense used here is an event where the proton intensity at energies above 10 MeV exceeds 10 pfu.

Based on the UMASEP scheme the UMASEP-10mw tool was developed for predicting >10 MeV SEP events using microwave data, the time derivative of the SXR was replaced by the microwave flux density. The UMASEP thresholds were re-calibrated. The tool UMASEP-10mw has been developed to be used for calculating the correlation between the solar microwave flux densities at 5 GHz and 8.8 GHz, which are monitored by patrol instruments, and the time derivatives of the near-earth differential proton fluxes measured in different energy channels (i.e. using the GOES satellites).

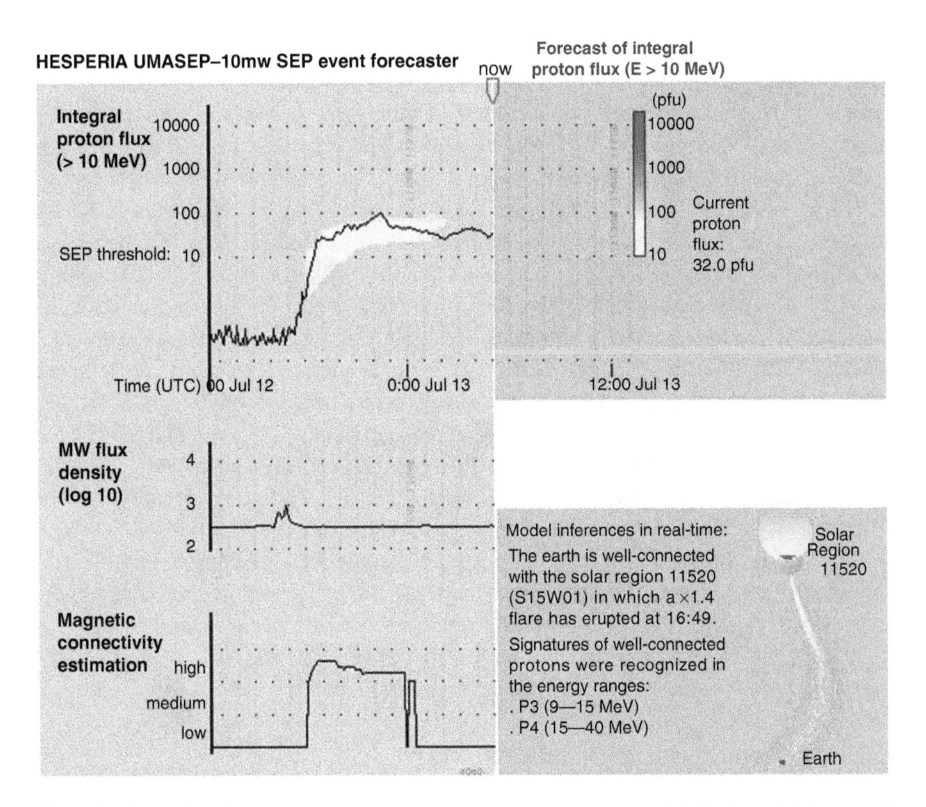

Fig. 7.2 UMASEP-10mw output after processing microwave data at 5 GHz from 2012 July 12 and GOES proton fluxes of >10 MeV energies. The *yellow/orange* band in the proton intensity plots gives the predicted range, with the colour scale shown by the *vertical bar*

We illustrate the forecast of the UMASEP-10mw tool using microwave data at 5 GHz for predicting the >10 MeV SEP event.

Figure 7.2 shows the forecast graphical output that an operator would have seen if the UMASEP-10mw tool had processed real-time microwave data on 2012 July 12. The upper time series in both images shows the observed integral proton flux with energies greater than 10 MeV. The current flux is indicated below the label "now" at each image. To the right of this label, the forecast integral proton flux is presented as a yellow/orange-coloured band. The central curve in each panel displays the microwave flux density time profile, and the lower time series shows the magnetic connectivity estimation with the best-connected coronal mass ejection (CME)/flare process zone.

Figure 7.2 also shows the prediction at 18:05 (2012 July 12). This forecast is that an event will start during the following 2 h and reach a peak intensity of 36 pfu 9 (see white section "Automatic forecast"). Below the forecast section, the system also presents the model inference section, which shows that the Earth is well-connected with the solar region 11520. The system also shows that the associated X1.4 flare took place at S15W01. As time passes, the integral proton flux also rises. At 18:35 UT, the flux exceeds the 10-pfu threshold, which indicates that a proton event is occurring. Note that the well-connected SEP event was successfully forecast 30 min earlier, when the enhancement of the integral proton flux was still weak (1.24 pfu).

To measure the overall performance of this tool, we run the UMASEP forecasting schemes using on the one hand the SXR observations, on the other hand the microwave observations at the two frequencies considered for the aforementioned period. We evaluated two quantities: the probability of detection (POD) is the number of the predicted SEP events divided by that of the SEP events that actually occurred, i.e. nine events in the considered time interval. The false-alarm ratio (FAR) is the number of false predictions over the number of predictions. Seven predictions were triggered when microwaves were used, and eight with SXR. We found that the use of SXR and microwave data produced the same POD = 77.8% (7/9). The most notable difference is that the use of microwave data does not yield any false alarm. The average warning time (AWT) is slightly higher when microwave observations are used 30.7 min as compared to 26.4 min.

The probabilities of detection used above are adequate to compare the performance of SXR and microwaves within the UMASEP scheme, but overestimate the expected ones: SEP events originating behind the solar limb are undetectable to the UMASEP Well-Connected Prediction model (WCP), because it uses electromagnetic observations from a terrestrial vantage point. This bias affects SXR from GOES and radio observations from ground alike.

A more detailed account of this work is given in Zucca et al. (2017).

7.3 Predicting SEP Energy Spectra from Historical Microwave Data

Depending on their peak intensity and spectral hardness, SEP events constitute different kinds of space weather hazard. Protons and heavy ions at energies between several MeV and several tens of MeV may interact with spacecraft and human beings above low-Earth orbit, and ionize the high polar atmosphere of the Earth. GeV protons create atmospheric cascades down to the Earth and enhanced radiation doses at aircraft altitudes. High intensities in the two energy ranges are not necessarily observed in the same events (Mewaldt et al. 2007). Besides the occurrence, spectral hardness is therefore a space-weather relevant information, and a second goal in SEP forecasting.

It was shown in Grechnev et al. (2015 and references therein) that SEP events above 100 MeV are often accompanied by strong microwave emission well above the average peak frequency of 10 GHz. On the other hand the peak flux density or peak fluence of microwave bursts also show some correlation with the peak intensity of SEPs at tens of MeV (Kahler 1982; Trottet et al. 2015). Chertok et al. (2009) went one step further and suggested that SEP events with hard proton

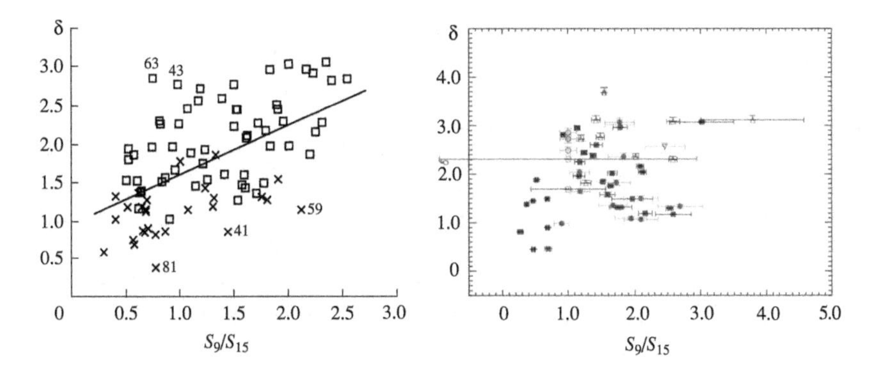

Fig. 7.3 Scatter plots of the spectral hardness of the proton spectrum δ, versus the peak microwave flux density ratio at 8.8 and 15.4 GHz, labelled S9/S15. (**a**) Study by Chertok et al. (2009). (**b**) Present work based on data from GOES integral intensities above 10 MeV and above 100 MeV taken during solar cycles 23 and 24

spectra in space in the 10–100 MeV range tend to be accompanied by microwave bursts where the flux density at the highest frequency continually monitored from ground, which is 15.4 GHz, exceeds the flux density at 9 GHz. The ratio of flux densities at the two frequencies, which is an easily observable parameter, seems to correlate significantly ($r = 0.55$) with the proton spectral hardness during solar cycles 22–23.

We re-examined the relationship during cycles 23 and 24, using the integral proton intensities measured by the GOES. We consider integral intensities above 10 MeV (designated by J_{10} in the following) and 100 MeV (J_{100}) for events associated with activity in the western solar hemisphere, and calculated the ratio

$$\delta = \log_{10}\left(\frac{J_{10}}{J_{100}}\right) \tag{7.1}$$

In Fig. 7.3 the scatter plot between the proton spectral hardness δ and the ratio of peak flux densities at 8.8 and 15.4 GHz derived by Chertok et al. (2009) (a) is compared with our work (b). Both plots suggest a slight trend that SEP events with hard proton spectra are associated with microwave bursts that are stronger at 15.4 than at 8.8 GHz. But the correlation is questionable in cycles 23–24: the linear correlation coefficient of the sample in Fig. 7.3b is 0.26 ± 0.20 in solar cycle 23, and still weaker in solar cycle 24. So we find no convincing correlation that could support a forecasting procedure of spectral hardness of SEPs. We also tested a correlation between spectral hardness and the speed of the associated CME. This was also inconclusive, with a correlation coefficient of 0.15 ± 0.16 in solar cycle 23.

There is a number of reasons why a relationship between the microwave peak frequency and the SEP spectral hardness in the range (10–100) MeV could be masked. One is the likely contribution of different acceleration processes to the SEP populations (Trottet et al. 2015) and the variation of their contribution with particle energy (Dierckxsens et al. 2015). If the SEPs above 10 MeV were predominantly accelerated by CME shocks, and those above 100 MeV by flares or similar processes lower in the corona, no direct correlation would be expected. The other reason is that the microwave flux density spectrum depends strongly on the magnetic field strength and orientation in the radio source, which is also not expected to have an effect on the SEPs. Finally there is an interesting hint that radio bursts with relatively high flux density at 15.4 GHz (and flat SEP spectra) were lacking in solar cycle 24. Forecasting schemes can of course use empirical correlations independently of our understanding of the physical relationships. But this does not seem convincing in the present case.

7.4 Predicting 30–50 MeV SEP Events by Using the RELeASE Scheme

The fact that near relativistic electrons (1 MeV electrons have 95% of the speed of light) travel faster than ions (30 MeV protons have 25% of the speed of light) and are always present in SEP events, a forecast of the arrival of protons from SEP events can be based on real-time measurements of near relativistic electrons. The faster electrons arrive 30–90 min before the slower protons at Lagrangian point 1. The Relativistic Electron Alert System for Exploration (REleASE) forecasting scheme uses this effect to predict the proton flux by utilizing the actual electron flux and the increase of the electron flux in the last 60 min. A detailed description of the REleASE scheme can be found in (Posner 2007). The original REleASE code uses real-time electron flux measurements from the Electron Proton Helium Instrument (EPHIN) (Müller-Mellin 1995) on board the Solar and Heliospheric Observatory (SOHO) to forecast the expected proton flux.

REleASE is based on electron intensity measurements of energies from 0.25 to 1 MeV and their intensity changes. It utilizes an empirical matrix in order to predict the proton intensity 30, 60 or 90 min ahead. Figure 7.4b displays the forecast matrix for one proton channel and the 60-min interval. EPHIN provides realtime data which are used with the REleASE scheme. One disadvantage of EPHIN data is the limited time coverage in the realtime data of less than 4 h per day. If no realtime data are available as input for the REleASE scheme, no forecast can be produced. The Electron Proton Alpha Monitor (EPAM) onboard the Advanced Composition Explorer (ACE) was selected to be a good candidate to deliver continuously input for REleASE because of the nearly full time coverage. EPAM provides realtime electron intensities in a comparable energy range (0.175–0.315 MeV vs. 0.25–1.0 MeV) but in a time resolution of 5 min instead of 1 min.

The forecast depends on the measured electron intensities and their increase. Hence we decided to determine a correlation between the intensities and the increase parameter. Figure 7.4a shows the time profile of EPAM (red) and EPHIN

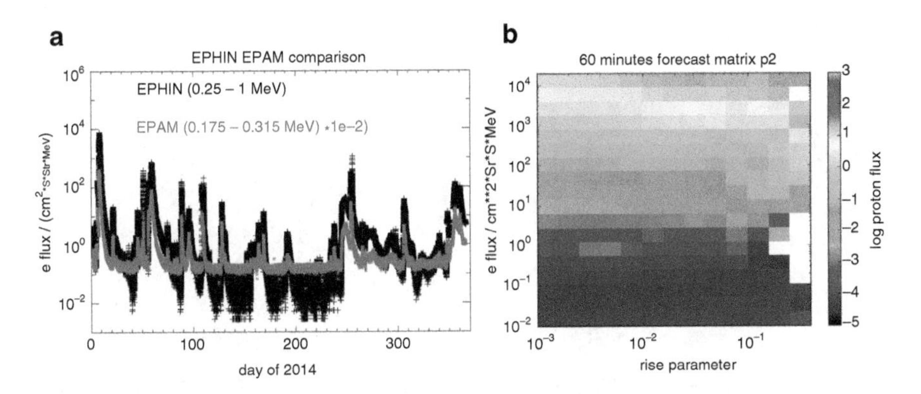

Fig. 7.4 The *left* panel (**a**) shows the time series of EPHIN (*black*) and EPAM (*red*) electron intensities. The electron intensity measured by EPAM is divided by 10. The *right* panel (**b**) displays an example of one forecast matrix used within the REleASE scheme. This Matrix shows the predicted intensity of protons in 1 h as function of the measured absolute electron intensities and the intensity rise parameter

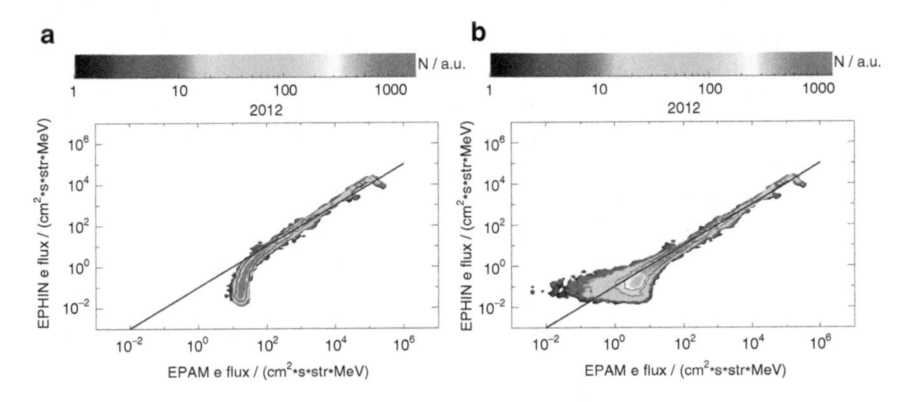

Fig. 7.5 The *left* panel (**a**) displays 5 min electron intensities of EPAM (x-axis) and EPHIN (y-axis) plotted against each other. The *right* panel (**b**) shows the same plot for background subtracted data

(black) electron intensities of 2014. Despite the high background of the EPAM measurements there seems to be a good correlation. If there is an increase in EPHIN electron intensity there is also one in EPAM.

To quantify this correlation Fig. 7.5a shows the EPHIN electron intensity on the y-axis plotted against the corresponding EPAM intensities on the x-axis. The higher background level of EPAM reflects itself by the nearly vertical line at low EPAM intensities. In order to correct for that we subtracted a background intensity of 18 $(\text{cm}^2\ \text{s sr MeV})^{-1}$ from the EPAM data. The result of this procedure is shown in Fig. 7.5b in the right panel. Despite these differences the EPAM electron intensity is roughly ten times higher as the one determined by EPHIN. This is indicated by the black line showing the function where EPAM intensity is ten times higher than EPHIN intensities.

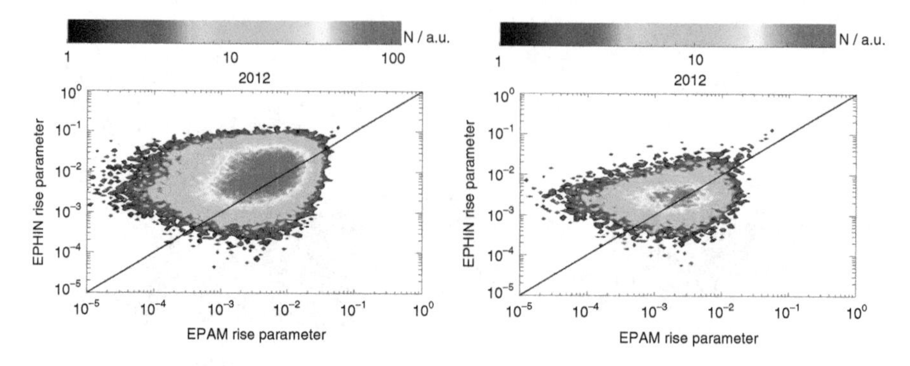

Fig. 7.6 Rise parameter of 5 min electron intensities measured by EPAM (x-axis) and EPHIN (y-axis). The EPAM intensity was background subtracted. The *right* panel displays all data from 2012 while the *left* panel shows only rise-parameters for strong enhanced electron fluxes

From our investigation, we conclude that there is a good correlation for intensities higher than 20 $(\text{cm}^2\ \text{s sr MeV})^{-1}$. Due to the large background the correlation breaks down for EPAM intensities below 20 particles $(\text{cm}^2\ \text{s sr MeV})^{-1}$. Since the goal of the REleASE system is to forecast SEP events with high particle fluxes the background uncertainties play a negligible role here. The EPAM intensity in addition was raised to the power of 1.02 to correct for different correlation at very high intensities.

The second parameter used in the REleASE forecast matrices (x-axis) is the intensity rise parameter. This parameter is calculated by linear fits through the logarithmic electron intensities of the last 5–60 min. The maximum of these parameters is transferred to the rise parameter position in the forecast matrices. A comparison of the rise parameters from EPHIN and EPAM is shown in Fig. 7.6. One can see that most of the data points are close to the bisecting line, but do not show a correlation between the rise parameters. If we only take electron fluxes higher than 102 into account, the correlation gets stronger. Due to this and the fact that the forecast matrix in Fig. 7.4 shows a much stronger dependence on the electron intensity we decided to use the uncorrected EPAM rise parameter as input to the REleASE scheme. Figure 7.7 shows an example of an SEP event where the EPHIN based and EPAM based forecasts predicted the real proton flux very accurately.

Since forecasts are made for different time offsets (30, 60, 90 min) and different overlapping energy channels (16–40 and 28–50 MeV), we investigated a suitable forecast condition in combining the different forecasts. We tested different combination of forecasts for different time offsets and found that the following condition delivered the best performance of the forecast systems:

- Alarm: if any forecast $> 10^{-1}$ $(\text{cm}^2\ \text{s sr MeV})^{-1}$ and one 30 min forecast $> 10^{-2}$ $(\text{cm}^2\ \text{s sr MeV})^{-1}$.
- Event: if real proton flux $> 10^{-1}$ $(\text{cm}^2\ \text{s sr MeV})^{-1}$.

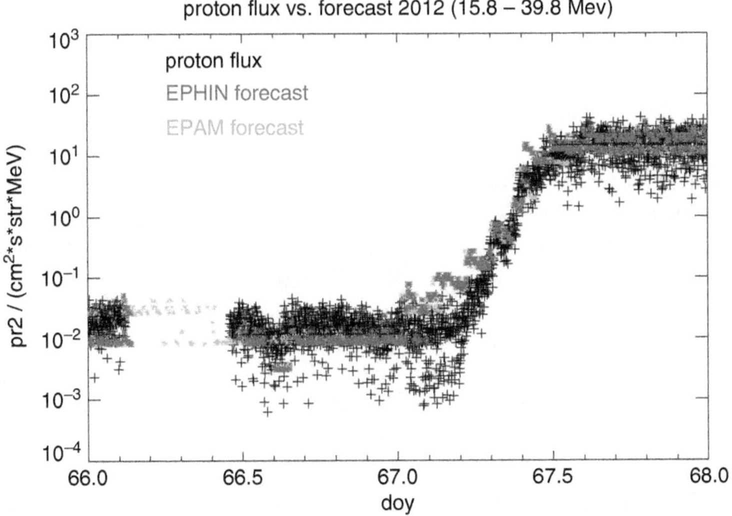

Fig. 7.7 An example of an SEP event where the EPHIN and EPAM based forecasts predicted the real proton flux accurately

Table 7.1 Results of REleASE implementation utilizing either SOHO/EPHIN or ACE/EPAM

	T	M	F	POD (%)	FAR (%)	AWT (min)
EPHIN	24	14	10	63	29	107
EPAM	24	14	13	63	35	123

The performance was tested by using all historic EPAM and EPHIN data from 2009 until 2016. All SEP events of this time period were investigated. The following results for events are possible:

- True forecast (Alarm and Event): T
- Missed event (No/late Alarm and Event): M
- False alarm (Alarm and no Event): F

By using the total number of true forecasts, missed events and false alarms of the analysed time period, it is possible to calculate the Probability of Detection (POD) and False Alarm Ratio (FAR):

- False Alarm Ratio: $FAR = \frac{F}{T+F}$
- Probability Of Detection: $POD = \frac{T}{T+M}$

The results of the described analysis are summarized in Table 7.1.

The described forecast tools have low FARs and sufficient PODs. The tools are publicly available via the HESPERIA project web site. On this web site we provide an e-mail alert system. Interested users are welcome to sign in for this alert system.

7.5 Predicting >500 MeV SEP Events by Using the UMASEP Scheme

Solar energetic particles (SEPs) are sometimes energetic enough and the flux is high enough to cause air showers in the stratosphere and in the troposphere, which are an important ionization source in the atmosphere. >500 MeV solar protons are so energetic that they usually have effects on the ground, producing what is called a Ground Level Enhancement (GLE). One of the goals of the HESPERIA project was the development of a predictor of >500 SEP events at the near-earth (e.g. at geostationary orbit). The implemented predictor, called HESPERIA UMASEP-500, makes a lag-correlation between the SXR flux and high-energy differential proton fluxes of the GOES satellites. When the correlation estimation surpasses a threshold, and the associated flare is greater than a specific SXR peak flux, a >500 MeV SEP forecast is issued.

The lag-correlation is carried out using the High-Energy UMASEP approach explained in Núñez (2015). In this project, this approach uses 1-min SXR and proton data. Firstly, it generates a bit-based time series from the SXR time-derivatives and three bit-based time series from the time-derivatives of each of the P9–P11 channels of the GOES6-GOES15 satellites. The “1s” in each bit-based time series are set when its positive time derivative surpasses a percentage *p* of the maximum value of the time derivative in the present sequence of size *L* (beyond which no influence is assumed in the SEP event to be predicted); otherwise, the flux level is transformed into a “0”. To avoid false alarms due to relatively strong fluctuations during periods of low solar activity, a threshold *d* is necessary, which is the minimum value to consider it as positive fluctuation (i.e., a “1”). This forecasting approach creates a list of *cause-consequence pairs* as follows: it takes the first “1” of the SXR-based time series, and the first “1” of the proton-based time series, to create a *pair*; it then takes the second pair of “1s” in each time series, and thus successively, until all the “1s” of the SXR-based time series are inspected. After following this procedure, if a “1” does belong to any pair, it is classified as an “odd”. For each pair, the *pair separation* between the SXR-based “1” and the proton-based “1” is calculated.

An ideal magnetic connection is detected when a sequence of SXR-based “1s” in a row is followed by a sequence of proton-based “1s” in a row. In an ideal magnetically connected event, all pairs have the same temporal separation, and no odd “1” has been found; in other words, an ideal magnetic connection is detected when all recently-measured strongest rises in the SXR flux are followed, some minutes later (i.e. the lag), by all recently-measured strongest rises in a proton channel. We say that this ideal magnetic connection would have a *Fluctuation Correlation* of 1. In general, we need a formula, described in Núñez (2015), that calculates the Fluctuation Correlation between the bit-valued SXR-based time series and a proton-based time series. A >500 MeV SEP event is triggered when the lag-correlation is greater than a threshold *r*, and the SXR intensity of the associated flare is greater than a threshold *f*.

It is important to mention that a >500 MeV SEP event is detected when the integral proton flux surpasses a certain threshold *pfu*500. To calculate this threshold, firstly we had to use the geometrical factors of P9, P10 and P11 proton channels provided by the National Oceanic and Atmospheric Administration (NOAA). Then, we manually varied this threshold to match each >500 MeV SEP event with each GLE event. The study ended with a threshold of 0.8 pfu, which yielded a one-to-one correspondence in 26 of the events of all 32 GLE events within the analyzed period (1986–2016). In 8 cases, a GLE was observed at Earth; however, the enhancement in >500 MeV integral proton flux did not surpass 0.8 pfu. In only one case (see Event 44.5 in Table 7.2) a >500 MeV SEP event took place, which was not observed at the ground.

The UMASEP-500 model's parameter calibration from historical data was an optimization process whose purpose was to obtain a high POD and Advance Warning Time (AWT), and a low FAR. We found that the same thresholds and parameters found for predicting >500 MeV SEP events were also very appropriate to predict GLE events; for this reason, this section also presents a summary of the GLE forecast results. For more information about the GLE forecast results, please consult Núñez et al. 2017.

The original purpose of the HESPERIA UMASEP-500 tool was to correlate SXR with neutron and proton data. We found that the use of proton data alone allowed this tool to make predictions before any Neutron Monitor (NM) station's alert. This satisfactory result became our operational criterion for classifying a GLE forecast as successful. We found that the correlation of SXR and neutron counter data did not trigger any hit additional to those generated using proton data alone. We also found that the use of neutron data provoked the generation of many false alarms due to some quality data problems (mainly spikes) caused by technical issues, such as problems in the neutron sensor tubes and power supplies, among others (Souvatzoglou et al. 2014). Since the use of neutron data did not increment the POD, but did increment the FAR, we decided not to use neutron data for making predictions.

Figure 7.8 presents the forecast output for the >500 MeV and GLE event on October 28, 2003. The upper time series shows the recent >500 MeV proton flux; the predicted flux is presented with a colored curve. The middle time series shows the recent SXR flux. The plot at the bottom of the forecast output presents the empirically-estimated level of magnetic connectivity. The Automatic Prediction section (on the right) presents in red the prediction of the occurrence of the GLE event. Below, the tool presents the details of the associated flare, and the proton channels for which the SXR correlation was found. The small image at the top-right shows the real evolution of the integral proton flux after this prediction. The 1-min real-time forecast outputs of this tool are shown on the website of the HESPERIA project (i.e. https://www.hesperia.astro.noa.gr/index.php/results/real-time-prediction-tools/umasep).

The overall prediction performance of event occurrences for the analyzed period (1986–2016) was calculated in terms of POD, FAR and AWT. Table 7.2 presents the list of GLEs, >500 MeV SEP events, and the HESPERIA UMASEP-500's

Table 7.2 HESPERIA UMASEP-500's forecast results of the GLE and >500 MeV SEP events that took place during the analyzed period (1986–2016)

Solar event Event ID	GLE Onset (UTC)	Time	GLE Forecast Result[a]	GLE Forecast Warning Time (min)[a]	>500 SEP Event Onset Time (UTC)	>500 SEP Forecast Result	>500 SEP Forecast Warning Time (min)
GLE-40	25/07/1989	8:50	Miss		9:13	Miss	
GLE-41	16/08/1989	1:25	Hit	20	1:37	Hit	32
GLE-42	29/09/1989	11:40	Hit	6	11:51	Hit	17
GLE-43	19/10/1989	13:00	Hit	8	13:03	Hit	11
GLE-44	22/10/1989	17:55	Miss		17:57	Miss	
Ev-44.5	23/10/1989				12:42	Miss	
GLE-45	24/10/1989	18:18	Miss		18:24	Miss	
GLE-46	15/11/1989	7:00	Miss				
GLE47	21/05/1990	22:29	Miss		23:02	Miss	
GLE-48	24/05/1990	20:49	Miss		21:06	Miss	
GLE-49	26/05/1990	20:55	Miss		21:10	Miss	
GLE-50	28/05/1990	5:34	Miss				
GLE-51	11/06/1991	2:30	Hit	24	3:18	Hit	72
GLE-52	15/06/1991	8:35	Hit	2	8:40	Hit	7
GLE-53	25/06/1992	20:15	Miss		20:18	Miss	
GLE-54	02/11/1992	3:50	Hit	20	3:24	Hit	28
GLE-55	06/11/1997	12:10	Miss		12:27	Miss	
GLE-56	02/05/1998	13:55	Miss				
GLE-57	06/05/1998	8:25	Miss				
GLE-58	24/08/1998	22:50	Miss				
GLE-59	14/07/2000	10:34	Hit	12	10:36	Hit	14
GLE-60	15/04/2001	13:57	Hit	1	13:57	Hit	1

GLE-61	18/04/2001	2:33	Miss				
GLE-62	04/11/2001	16:55	Miss		16:44	Miss	
GLE-63	26/12/2001	5:39	Miss		6:08	Miss	
GLE-64	24/08/2002	1:23	Hit	13	1:30	Hit	20
GLE-65	28/10/2003	11:17	Hit	9	11:31	Hit	23
GLE-66	29/10/2003	21:02	Hit	3	21:19	Hit	20
GLE-67	02/11/2003	17:27	Hit	3	17:35	Hit	11
GLE-68	17/01/2005	9:52	Miss				
GLE-69	20/01/2005	6:47	Miss		6:46	Miss	
GLE-70	13/12/2006	2:50	Hit	15	2:59	Hit	24
GLE-71	17/05/2012	1:55	Miss		2:07	Miss	

[a]Hits are predictions that are triggered before the first NM station's GLE alert

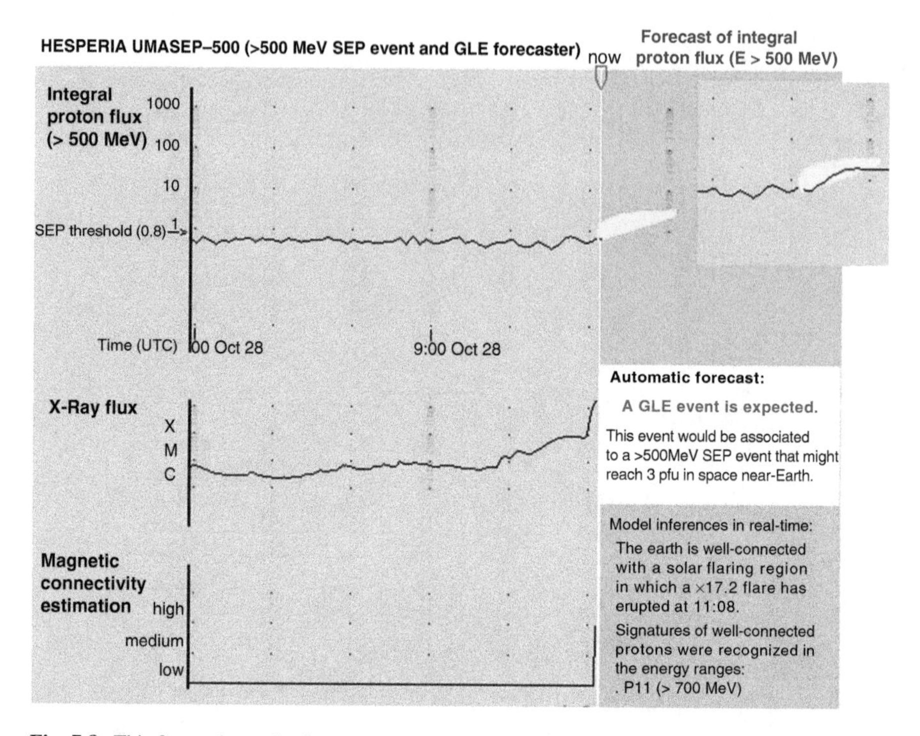

Fig. 7.8 This figure shows the forecast output of HESPERIA UMASEP-500 for the event occurred on October 28, 2003. This prediction, issued at 11:08, was successful because it was issued before the >500 MeV integral proton flux surpassed 0.8 pfu (at 11:31), and before the first NM station's alert, issued at 11:17

prediction results for the events of the analyzed period. Column 1 presents the GLE event ID; column 2 lists the time of the first detection of the event of an NM station; column 3 presents the HESPERIA UMASEP-500 GLE prediction results (hits are those events successfully predicted, and misses are those not-successfully predicted); column 4 lists the warning times (i.e. the temporal difference between the time at which the forecast was triggered by this tool and the time of the first NM station's alert); column 5 lists the time the occurrence of the SEP event (i.e. the time when the >500 MeV proton flux surpassed 0.8 pfu); column 6 presents the SEP prediction results; and, column 7 lists the warning times.

Regarding the prediction of >500 MeV SEP event, the forecast performance results of this tool for the period 1986–2016 may be summarized as follows: the POD was 50% (13/26), the FAR was 31.6% (6/19), and the average warning time to the first NM's alert was 13 min.

Regarding the prediction of GLE events, the forecasting results for the most recent half of the evaluation period (i.e. 2000–2016) may be used to compare UMASEP-500 with those of the GLE Alert Plus. These results may be summarized as follows: the POD was 53.8% (7 of 13 GLE events); the FAR was 30.0% (3/10);

the AWT to the first NM's alert was 8 min; and, the AWT to the GLE Alert Plus's warning was 15 min. The GLE forecasting results for the first half of the evaluation period (i.e. 1986–1999) are summarized as follows: the POD was 31.6% (6 of 19 GLE events); the FAR was 33.3% (3/9); and, the AWT to the first NM's alert was 13.3 min. There are no forecasting results of the GLE Alert Plus for this period.

For the whole evaluation period, the GLE forecasting performance may be summarized as follows: the POD was 40.6% (13 of 32); the FAR was 31.6% (6 of 19); and, the AWT to the first NM's alert was 10.5 min. Note that the FAR of the most recent period is similar to that of the oldest period (30.0% vs. 33.3%); however, the POD of the most recent period (i.e. 53.8%) is better than the POD of the oldest period (i.e. 31.6%). We do not know the reason for the better POD performance in the most recent period; nevertheless, we think that the use of a more recent and refined instrument technology and/or more experienced calibration procedures yields better forecasting performance.

7.6 Concluding Remarks

We experimented the use of microwave time histories in the UMASEP prediction scheme of the occurrence of SEP events. The test run over 13 months shows that microwaves provide a comparable probability of detection, but a reduced false-alarm ratio as compared to the time derivative of the SXR flux, which is used in the traditional UMASEP scheme.

The reduction of false alarms is due to the fact that microwave bursts are signature of non-thermal particle acceleration and are less frequent than the ubiquitous thermal soft X-ray brightenings. This reduces the probability to interpret the chance coincidence between a rise of the radiative signature and the rise of the particle intensity at the spacecraft as an indication of a magnetic connection. The forecasting scheme using microwaves fails when the microwave emission is thermal and slowly rising. This is especially the case when SEP events are related to the eruption of quiescent filaments.

A second test of microwave patrol observations in SEP forecasting was conducted with the aim to predict the hardness of proton spectra using the ratio of peak flux densities at 15.4 and 8.8 GHz: the expectation was to find a preferential association of hard proton spectra with microwave bursts that are particularly strong at 15.4 GHz, as had been shown in previous activity cycles (Chertok et al. 2009). We were unable to confirm this expectation: we found no significant correlation between the proton spectral hardness and the microwave flux density ratio. The intrinsic variations from event to event are much stronger than any underlying trend that might exist.

The radio patrol observations used by our study are carried out with rather simple patrol instruments, which monitor the whole Sun flux density using parabolic antennas with a typical size of $\sim$1 m. Such data are presently not provided in real time, but there is no technical obstacle to do so. But the results of our test run for the prediction of well-connected SEP events show that microwave observations have the potential to improve SEP forecasting. An interesting perspective could be the combination of the REleASE and UMASEP forecasting schemes, because, on one hand they could correlate rises between microwaves at the Sun and electrons, and on the other hand, forecasts may be provided for those SEPs whose parent solar event is behind the limb. This combination of schemes could bring a major gain in advance warning time.

The HESPERIA REleASE tools make real-time predictions of the proton flux (30–50 MeV) at Lagrangian point 1 and are available via the HESPERIA web site.[3] An analysis of historic data from 2009 to 2016 has shown that the forecast tools have a low FAR (∼30%) and a high POD (63%).

The HESPERIA UMASEP-500 model makes real-time predictions of the occurrence of >500 MeV SEP and GLE events from the analysis of SXR and differential proton flux measured by the GOES satellite network. Real-time predictions are available in the HESPERIA web site.[4] We assume that a prediction is successful when it is reported before the first GLE alert is issued by any NM station. Regarding the prediction of GLE events for the period 2000–2016, this tool had a POD of 53.8%, and a FAR of 30.0%. For this period, the tool obtained an AWT of 8 min taking as reference the alert time from the first NM station; taking as reference the time of the warnings issued by the GLE Alert Plus for the aforementioned period, the HESPERIA UMASEP-500 tool obtained an AWT of 15 min.

In summary, the goal of the presented tools has been to improve mitigation of adverse effects both in space and in the air from a significant solar radiation storm, providing valuable added minutes of forewarning to space weather users.

References

Chertok, I., Grechnev, V., Meshalkina, N.: On the correlation between spectra of solar microwave bursts and proton fluxes near the Earth. Astron. Rep. **53**, 1059–1069 (2009)

Dennis, B.R., Zarro, D.: The neupert effect – what can it tell us about the impulsive and gradual phases of solar flares? Sol. Phys. **146**, 177–190 (1993)

Dierckxsens, M., Tziotziou, K., Dalla, S., Patsou, I., Marsh, M., Crosby, N.: Relationship between solar energetic particles and properties of flares and CMEs: statistical analysis of solar cycle 23 events. Sol. Phys. **290**, 841–874 (2015)

Grechnev, V., Kiselev, V., Meshalkina, N., Chertok, I.: Relations between microwave bursts and near-earth high-energy proton enhancements and their origin. Sol. Phys. **290**, 2827 (2015)

Holman, G., Aschwanden, H., Aurass, M., Tanaka, S.: Implications of X-ray observations for electron acceleration and propagation in solar flares. Space Science. **159**, 107–166 (2011)

Kahler, S.: The role of the big flare syndrome in correlations of solar energetic proton fluxes and associated microwave burst parameters. Space Phys. **87**, 3439–3448 (1982)

[3]https://www.hesperia.astro.noa.gr/index.php/results/real-time-prediction-tools/release

[4]https://www.hesperia.astro.noa.gr/index.php/results/real-time-prediction-tools/umasep

Mewaldt, R., Cohen, C., Mason, G., Cummings, A., Desai, M., Leske, R., Raines, J., Stone, E., Wiedenberck, N., von Rosenvinge, T., Zurbenchen, T.: On the differences in composition between solar energetic particles and solar wind. Space Sci. Rev. **130**, 207–219 (2007)

Müller-Mellin, R.: COSTEP – comprehensive suprathermal and energetic particle analyser. Sol. Phys. **162**, 483–504 (1995)

Nakajima, H., Sekiguchi, H., Sawa, M., Kai, K., Kawashima, S.: The radiometer and polarimeters at 80, 35 and 17 GHz for solar observations at Nobeyama. Publ. Astron. Soc. J. **37**, 163–170 (1985)

Neupert, W.: Comparison of solar X-ray line emission with microwave emission during flares. Astrophys. J. **153**, L59–L64 (1968)

Núñez, M.: Predicting solar energetic proton events (E>10 MeV). Space Weather. **9**, S07003 (2011)

Núñez, M.: Real-time prediction of the occurrence and intensity of the first hours of >100 MeV solar energetic proton events. Space Weather. **13**(11), 807–819 (2015)

Núñez, M., Reyes-Santiago, P., Malandraki, O.E.: Real-time prediction of the occurrence of GLE events. Space Weather 15 (2017). doi:10.1002/2017SW001605

Posner, A.: Up to 1-hour forecasting of radiation hazards from solar energetic ion events with relativistic electrons. Space Weather. **5**, S05001 (2007)

Souvatzoglou, G., Papaioannou, A., Mavromichalaki, H., Dimitroulakos, J., Sarlanis, C.: Optimizing the real-time ground level enhancement alert system based on neutron monitor measurements: introducing GLE Alert Plus. Space Weather. **12**(11), 633–649 (2014)

Torii, C., Tsukiji, Y., Kobayashi, S., Yoshimi, N., Tanaka, H., Enome, S.: Full-automatic radiopolarimeters for solar patrol at microwave frequencies. Proc. Res. Inst. Atmos. **26**, 129–132 (1979)

Trottet, G., Samwel, S., Klein, K., Dudok de Wit, T., Miteva, R.: Statistical evidence for contributions of flares and coronal mass ejections to major solar energetic particle events. Sol. Phys. **290**(3), 819–839 (2015)

Zucca, P., Núñez, M., Klein, K.: Exploring the potential of microwave diagnostics in SEP forecasting: I. The occurrence of SEP events. J. Space Weather Space Clim. **7**, 15 (2017)

Chapter 8
X-Ray, Radio and SEP Observations of Relativistic Gamma-Ray Events

Karl-Ludwig Klein, Kostas Tziotziou, Pietro Zucca, Eino Valtonen, Nicole Vilmer, Olga E. Malandraki, Clarisse Hamadache, Bernd Heber, and Jürgen Kiener

Abstract The rather frequent occurrence, and sometimes long duration, of γ-ray events at photon energies above 100 MeV challenges our understanding of particle acceleration processes at the Sun. The emission is ascribed to pion-decay photons due to protons with energies above 300 MeV. We study the X-ray and radio emissions and the solar energetic particles (SEPs) in space for a set of 25 *Fermi* γ-ray events. They are accompanied by strong SEP events, including, in most cases where the parent activity is well-connected, protons above 300 MeV. Signatures of energetic electron acceleration in the corona accompany the impulsive and early post-impulsive γ-ray emission. γ-ray emission lasting several hours accompanies in general the decay phase of long-lasting soft X-ray bursts and decametric-to-kilometric type II bursts. We discuss the impact of these results on the origin of the γ-ray events.

K.-L. Klein (✉) • N. Vilmer • P. Zucca
LESIA-Observatoire de Paris, CNRS, 92190 Meudon, France

PSL Research University, Universités P & M. Curie, Paris-Diderot, Meudon, France
e-mail: ludwig.klein@obspm.fr; pietro.zucca@obspm.fr; nicole.vilmer@obspm.fr

K. Tziotziou • O.E. Malandraki
National Observatory of Athens, IAASARS, Athens, Greece
e-mail: kostas@noa.gr; omaland@astro.noa.gr

E. Valtonen
Department of Physics and Astronomy, Space Research Laboratory, University of Turku, Turku, Finland
e-mail: eino.valtonen@utu.fi

C. Hamadache • J. Kiener
CSNSM, IN2P3-CNRS, Univ. Paris-Sud, 91405 Orsay Cedex, France
e-mail: clarisse.hamadache@csnsm.in2p3.fr; Jurgen.kiener@csnsm.in2p3.fr

B. Heber
Christian-Albrechts-Universität zu Kiel, Kiel, Germany
e-mail: heber@physik.uni-kiel.de

O.E. Malandraki, N.B. Crosby (eds.), *Solar Particle Radiation Storms Forecasting and Analysis, The HESPERIA HORIZON 2020 Project and Beyond*, Astrophysics and Space Science Library 444, DOI 10.1007/978-3-319-60051-2_8

8.1 Introduction

The advent of the *Fermi* mission showed that the Sun is an occasional, but unexpectedly frequent, emitter of γ-ray photons above 100 MeV. These are understood to be produced by pion decays in nuclear interactions involving protons or He-nuclei at energies above 300 MeV/nucleon. One did not expect that the Sun was able to accelerate relativistic protons and nuclei even in seemingly modest flares. These particles are rarely detected in space (<1 event per year). Furthermore, the duration, several hours, of some γ-ray events is much longer than that of hard X-ray signatures of electron acceleration in the impulsive flare phase.

The question of how the γ-ray emission is related to other signatures of particle acceleration and energy release in the corona is crucial to understanding the origin of the high-energy protons. It might also be expected that such high-energy populations interacting at the Sun are accompanied by particularly energetic solar energetic particle (SEP) events. This chapter is based on 25 events. The *Fermi*/LAT temporal data were made available to the HESPERIA project by G. Share prior to their publication in a comprehensive paper (Share et al. 2017). The present chapter introduces the relevant process of emission and pre-*Fermi* observations of pion-decay γ-rays (Sect. 8.2), and gives an overview of the *Fermi*/LAT observations (Sect. 8.3). Section 8.3 was prepared by Gerald Share and Ron Murphy. Related X-ray and radio observations and associated SEP events are presented in Sects. 8.4 and 8.5, respectively. Preliminary conclusions on the interpretation of the γ-ray events are in Sect. 8.6.

8.2 Theory and Early Observations of Gamma-Ray Emission at Photon Energies >60 MeV

On 1982 Jun 3 the gamma-ray spectrometer on the *Solar Maximum Mission* satellite observed emission from 0.3 to 100 MeV from a X8.0 *GOES*-class flare (Forrest et al. 1986). The impulsive flare lasted about 1 min and was followed by a distinct harder emission phase that peaked in about 1 min and lasted for over 15 min. The energy spectrum of this sustained emission displayed a characteristic hump at photon energies above 60 MeV (Fig. 8.1a), which appeared to be consistent with that from pion-decays produced by the interaction of >300 MeV protons in the solar atmosphere (Forrest et al. 1985, see below); the authors speculated whether the emission might be associated with the acceleration of solar energetic particles. There were several more of these events in the ensuing years, all associated with intense X-class flares, that were summarized in a paper entitled "Long-Duration Solar Gamma-Ray Flares" (LDGRFs) (Ryan 2000). Various origins were suggested, and a

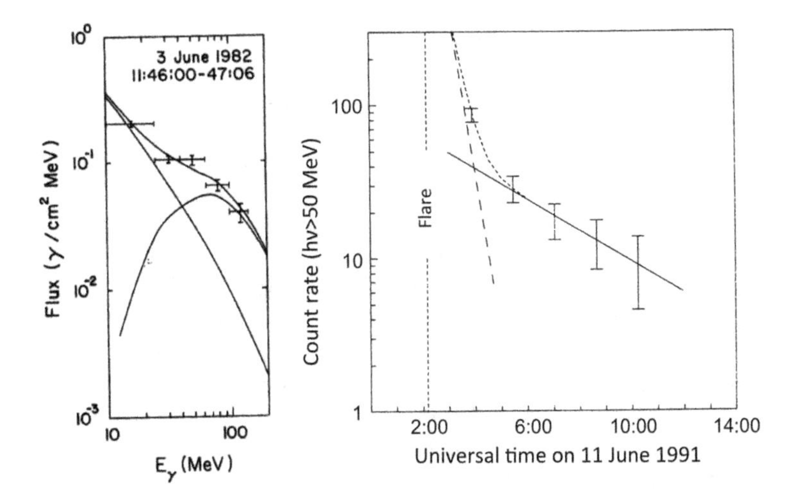

Fig. 8.1 (**a**) The γ-ray spectrum showing the pion-decay bump of the 1982 Jun 3 event, observed by SMM (Forrest et al. 1985). Credit: Forrest et al., Internat. Cosmic Ray Conf. 4, 146, 1985, courtesy W.T. Vestrand. (**b**) Time profile of the first sustained γ-ray event detected by the Compton Gamma-Ray Telescope (adapted from Kanbach et al. (1993); credit: Kanbach et al., A&A Suppl. 97, 349, 1993, reproduced with permission ©ESO)

key question was whether such emission would be observed when the accompanying flare was weaker and did not produce impulsive gamma radiation.

Detection of such events continued with observations from GAMMA-1 (Akimov et al. 1992), the Compton Gamma-Ray Observatory (CGRO) (Mandzhavidze and Ramaty 1992; Kanbach et al. 1993; Dunphy et al. 1999), GRANAT (Debrunner et al. 1997; Vilmer et al. 2003), CORONAS-F (Kuznetsov et al. 2011; Kuznetsov et al. 2014). Around 20 events had been observed then with significant emission above 60 MeV from pion decay radiation (see Lockwood et al. 1997; Chupp and Ryan 2009; Vilmer et al. 2011 for reviews). For some of the events, pion decay radiation is observed during the impulsive phase of the event as defined by the production of hard X-rays above 100 keV. In some events, high energy emissions had also been observed for hours after the impulsive phase of the flare, revealing that high energy ions are present on time scales of several hours (e.g. Kanbach et al. 1993; Ryan et al. 1994; Ryan 2000; Rank et al. 2001).

8.2.1 *Pion-Decay γ-Ray Emission*

High-energy (>60 MeV) emission in solar flares results from nuclear interactions of mildly relativistic ions (above a few hundred MeV/nucleon) with the ambient solar atmosphere (most probably the dense chromosphere). These nuclear reactions result in the production of pions. Charged pions decay to electrons and positrons, which produce γ-ray emissions through bremsstrahlung radiation. Energetic positrons also contribute to the γ-ray continuum by annihilating in flight. The neutral pions decay in two photons, one being emitted at high energies. The result is a very flat spectrum with a broad-bump feature with a maximum at 67 MeV (Murphy et al. 1987). In a magnetized region, synchrotron losses of electrons and positrons may be important and reduce their contributions with respect to the radiation from neutral-pion decay (Murphy et al. 1987; Vilmer et al. 2003). When energetic electrons are present at energies above 10 MeV, they produce bremsstrahlung emission above 10 MeV (Chap. 2), which may mask pion-decay radiation.

Early work on the modelling of pion decay radiation and γ-ray emission (produced by >10 MeV ions) (Murphy et al. 1997) gave a first determination of the high energy spectrum of ions in flares. Assuming a spectral shape for the energetic ions (either a Bessel function or a power law), number and spectra of energetic protons were estimated for both the impulsive and the extended phase of the first detected event (1980 Jun 21) and showed that the proton spectrum was steeper in the impulsive phase than in the later, so-called 'extended' phase. This evolution of the ion spectrum in the extended phase was confirmed by further observations such as the ones of the 1991 Jun 11 event by CGRO/EGRET (Dunphy et al. 1999).

Quantitative analysis of several events with significant pion production has been performed providing information on the ion energy spectrum above 300 MeV/nucleon and allowing a comparison of this spectrum with the one deduced at lower energies from γ-ray line spectroscopy (see, e.g., Alexander et al. 1994; Dunphy et al. 1999; Kocharov et al. 1994, 1998; Vilmer et al. 2003). These comparisons have shown that the ion energy distribution does not have a simple power-law form from the γ-ray line producing energy domain (1–10 MeV) to the pion-producing energy domain (>300 MeV/nucleon).

8.2.2 Long-Duration γ-Ray Events

Even before the *Fermi* era, a few events had been observed where enhanced pion-decay radiation lasted several hours, and the question of the origin of these long duration events had been examined. On 1991 Jun 11 (Fig. 8.1b) emission above 50 MeV was detected for almost 8 h after the flare by COMPTON/EGRET. Several interpretations had been proposed to explain these long duration emissions, either the continuous acceleration of protons above 300 MeV (e.g., Ryan and Lee 1991) or the trapping of protons on very long time-scales. In particular, Mandzhavidze and Ramaty (1992) showed that the long duration phase could be explained by the injection of energetic protons in the impulsive phase and subsequent trapping. An efficient trapping on such long timescales required a strong mirror ratio in the trapping region (>10) as well as a coronal density less than $5 \cdot 10^{11}$ cm^{-3}. The question of how trapped particle populations could remain stable over hours was, however, not explained satisfactorily. An observational justification of continuous time-extended acceleration came from the discovery of sustained γ-ray events of moderate duration (1–2 h), which were accompanied by non-thermal microwave emission (Kocharov et al. 1994; Trottet et al. 1994; Akimov et al. 1996).

8.3 New Insights of Sustained Emission Events from *Fermi* Observations

With the launch of the sensitive *Fermi* Large Area Telescope (LAT) (Atwood et al. 2009), it became possible to observe weak γ-ray emission from the Sun due to its large effective area and aperture, and excellent background rejection. This permitted detection of quiescent γ-ray emission from cosmic-ray protons interacting in the solar atmosphere and cosmic-ray electrons interacting with sunlight (Abdo et al. 2011) with fluxes of 5 and 7 $\times 10^{-7} \gamma$ cm^{-2} s^{-1}, respectively. LAT first detected transient >100 MeV solar γ-rays on 2010 Jun 12 from the impulsive M2 class flare (Ackermann et al. 2012a,b) during an $\sim$50 s period. The emission was delayed about 10 s from the associated hard X-ray and nuclear-line emission and there was no evidence for any >100 MeV γ-ray emission in the hours after the flare. The *Fermi*/LAT team reported the detection of 18 >100 MeV events associated with solar flares covering the time period from 2008 August to 2012 August (Ackermann et al. 2014) which they classified as being impulsive, sustained, or delayed. In some cases they categorized the events as having both impulsive and sustained characteristics. Only three of the events were classified as only having an impulsive component. We prefer to use the name "sustained" to categorize all emission that is distinct from the impulsive flare independent of its duration.

Details of the 2011 Mar 7 and 2011 Jun 7 γ-ray observations and related solar measurements were also presented in Ackermann et al. (2014). The March 7 event was reported as having both impulsive and sustained emission components, while the June 7 event was classified as only having sustained emission because γ-rays were detected in only one LAT exposure about 1 h after the flare and there was no exposure to the flare. The γ-ray spectra were fit by two empirical models, a power law and a power law with exponential cutoff; and a physical pion-decay spectrum based on Murphy et al. (1987). Only heuristic arguments were presented to justify that the observed spectrum was from pion decay and not from electron bremsstrahlung. There was clear evidence for spectral softening over the 13 h period that the March 7 event was observed. As LAT's instrumental point spread function is about 1°, even above 1 GeV, only the location of the centroid of the γ-ray source could be deduced. The centroids of both sources were consistent with the active regions with uncertainties on the scale of a solar quadrant.

Ajello et al. (2014) reported detection on 2012 Mar 7 by LAT of what appeared to be distinct impulsive and sustained-emission phases associated with X5.4 and X1.3 (actually about M7 when the tail of the X5 flare is subtracted) flares, CMEs with speeds of about 2700 and 1800 km s^{-1}, and a strong solar energetic particle event. The γ-ray flux was one of the brightest observed by LAT; it lasted close to 20 h and the fitted spectrum again softened with time. The time integrated centroid of the emission was consistent with the location of the flares with a 1 σ uncertainty of $\sim$10°. There is evidence that the source of the emission moved from the eastern to the western hemisphere about 7 h following the flares. Once again there was no information on the spatial extent of the emission.

With detection of >100 MeV γ-ray emission associated with two behind-the-limb solar flares, it has become clear that emission can extend as much as 40° from the flare site (Pesce-Rollins et al. 2015; Ackermann et al. 2017). The events occurred on 2013 Oct 11 and 2014 Sep 1 and were associated with fast CMEs and strong solar energetic particle events. The centroids of the γ-ray emission for both events were close to the solar limb. The centroid for the October 11 event was consistent with the N21 latitude of the active region, located about 13° behind the East limb, and its 1 σ uncertainty extended to about 50° East. In contrast, the centroid of the September 1 event was significantly north (~15°) of the flare site that was located about 35° behind the East limb.

More information on the characteristics and origin of these solar γ-ray events became available with the completion of a comprehensive study of 29 sustained emission events observed by LAT between 2008 and 2016 (Share et al. 2017). This study indicates that the emission is not spatially distributed globally on the Sun but is probably distributed over a few tens of degrees around the centroid location. From this study it is also clear that the sustained >100 MeV emission is temporally distinct from the impulsive flare phase, which often emits hard X-rays extending only to hundreds of keV. The time profile of the first event observed by LAT on 2011 Mar 7, discussed above and plotted in Fig. 8.2, provides an example of such a distinct sustained-emission phase. While the analysis done by Share et al. (2017) is different from that of Ackermann et al. (2014), it uses the same source-class data and yields comparable fluxes and spectral results. It is simply a 'light-bucket' in which photons arriving with directions <10° of the Sun are accumulated, as long as the Sun is far from the Earth's horizon. The main plot in Fig. 8.2 shows the hours-long time profile of this 13 h event with LAT solar exposures every 3 h. The background level contains roughly equal contributions from the quiescent Sun, and Galactic and extra-galactic γ rays. The emission began during the time of the *GOES* flare and rose to peak about 6 h later.

The inset in Fig. 8.2 shows >100 MeV fluxes plotted at 4 min resolution after the hard X-ray peak of the March 7 flare. For comparison, the dotted line follows the high-energy time profile observed in the 1991 June 11 event which exhibited both impulsive emission and sustained emission lasting 8 h. The >100 MeV emission on March 7 appears to increase during the observations after the impulsive phase. There is also no evidence for nuclear line emission during the impulsive flare, suggesting that >1 MeV protons were not present in significant numbers. A limit on the 2.223 MeV neutron capture line during the flare suggests that the flare had at most only 20% of the number of >500 MeV protons observed in the sustained emission phase (Share et al. 2017). This indicates that only sustained >100 MeV emission was observed by LAT on March 7, in contrast to what was reported by Ackermann et al. (2014). The same conclusion can be reached about other events reported by Ackermann et al. (2012b) to have impulsive >100 MeV phases. Specifically, Share et al. (2017) find that the first >100 MeV outburst on 2012 Mar 7 described in detail by Ackermann et al. (2014) was distinct sustained emission and not directly associated with the X5.4 flare.

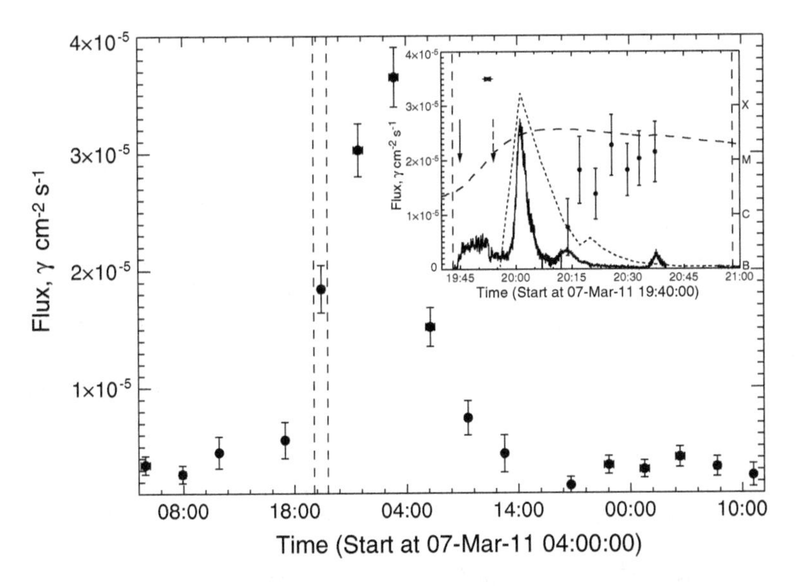

Fig. 8.2 Time history of the >100 MeV flux from $\leq 10°$ of the Sun revealing the 2011 Mar 7 LAT event (data points and $\pm 1\sigma$ statistical uncertainties). *Vertical dashed lines* show the *GOES* start and end times. The inset shows 4-min accumulation LAT data points and the merged and arbitrarily scaled >100 keV count rates observed by *RHESSI* and *Fermi*/GBM during the impulsive flare. The *dashed curve* shows the *GOES* 1–8Å profile (scale on right ordinate) and the < − > symbol shows the range in CME onset times in the CDAW catalog derived for linear and quadratic extrapolations. The *vertical solid arrow* depicts our estimate of the CME onset from inspection of *SDO*/AIA images and the *vertical dashed arrow* shows the estimated onset of Type II radio emission

Share et al. (2017) also find that 19 of the 29 sustained-emission events had time profiles distinct from the impulsive phase and most of these had onset times after the hard X-ray peaks. For the remaining events, there were not enough data to determine whether the sustained emission was distinct from or primarily associated with the impulsive phase, but in no case is there clear evidence that the sustained emission is the tail of the impulsive flare.

Pion-decay spectra from protons having different power-law spectral indices fit the spectra from the 2011 Mar 7 event and the other 28 events (Share et al. 2017). In general the results are consistent with those obtained by Ackermann et al. (2014, 2017) and Ajello et al. (2014). In addition, Share et al. (2017) demonstrated that only pion-decay spectra are consistent with the brightest sustained emission events and that any plausible electron bremsstrahlung spectra are not. Using nuclear line observations during these same bright sustained emission events, it was also possible to demonstrate that the proton spectra from 10 to 300 MeV were flatter than the spectra at higher energies. In addition to the two events found to have spectra that softened with time, Share et al. (2017) found two more that softened in time and two that hardened in time.

Detailed spectroscopic studies of many of the events observed by LAT revealed that the number of >500 MeV protons producing the sustained emission was typically at least a factor ten more than found in the accompanying impulsive flare. This is consistent with the distinctly different nature of the time profiles of the impulsive flares and the sustained emission discussed above. As Share et al. (2017) discuss, it is now clear that another energy source is necessary to accelerate protons to energies >300 MeV in order to produce the pion-decay emission observed in the sustained events. These energetic considerations and the rise in sustained γ-ray emission following the impulsive phases in many events suggests the likely source of the energy is the accompanying fast CME, possibly through its shock that is thought to produce gradual SEPs. The authors (Share et al. 2017) find that the number of >500 MeV SEP protons is on average about 100 times the number returning to the Sun to produce the sustained γ-ray emission. This is consistent with what shock wave models estimate (Kocharov et al. 2015).

The 2013 Oct 13 and 2014 Sep 1 behind-the-limb events (Ackermann et al. 2017) were also studied by Share et al. (2017). The time profile of the latter event, which lasted at least 2 h, is plotted in Fig. 8.3. Both >100 MeV γ-ray and >100 keV electron bremsstrahlung rose about 7 min following the impulsive hard X-ray flare, as inferred from SAX observations on MESSENGER (Schlemm et al. 2007). The bremsstrahlung extended up to at least 10 MeV and is consistent with production by

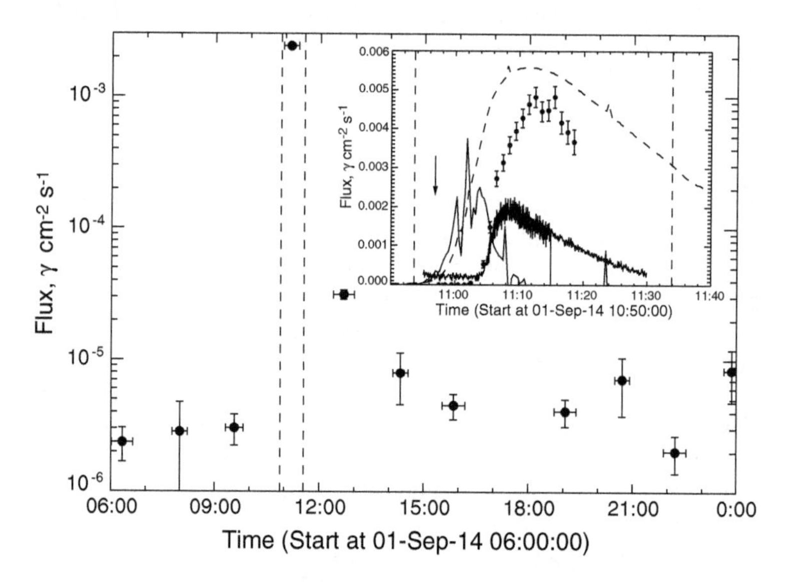

Fig. 8.3 Time history of the >100 MeV flux of the Sun revealing the 2014 Sep 1 LAT event (data points and $\pm 1\sigma$ statistical uncertainties). *Vertical dashed lines* show the soft X-ray start and end times from *MESSENGER*/SAX. The inset shows 1-min accumulation >100 MeV data points ($\pm 1\sigma$ statistical uncertainties) and the arbitrarily scaled 100–300 keV count rates observed by GBM. The *dashed curve* shows the soft X-ray profile from SAX and the *thin solid curve* its derivative. The *vertical solid arrow* depicts our estimate of the CME onset from inspection of *SDO*/AIA images

electrons interacting in a thick target, indicating the electrons were produced by the same CME shock that accelerated the protons producing the sustained γ radiation (Share et al. 2017). This conclusion is consistent with the finding that a direct magnetic connection exists between the shock wave and the low solar atmosphere at the onset of the hard X-rays and γ-rays in both behind the limb events (Plotnikov et al. 2017).

The conclusion reached by Share et al. (2017) is that the sustained emission events are likely due to shock-accelerated particles, associated with those in SEPs, that are imparted to field lines that return to the Sun and those that return to the Sun on open field lines from an SEP reservoir. This interpretation will be discussed in Sect. 8.4.

8.4 Multiwavelength Observations of *Fermi*/LAT γ-Ray Events

In this section we examine the relationship between the γ-ray events and other manifestations of energy release in the corona, namely heating as revealed by soft X-ray bursts, and electron acceleration traced by hard X-ray and radio emission. We use especially hard X-ray observations from RHESSI (Lin et al. 2002) and INTEGRAL/ACS (Rodríguez-Gasén et al. 2014), whole-Sun radio fluxes from the RSTN and Nobeyama Observatories (Torii et al. 1979; Nakajima et al. 1985) on ground, and from the radio spectrographs aboard Wind (Bougeret et al. 1995) and STEREO (Bougeret et al. 2008).

At decametric and longer waves (referred to as DH in the following) all 25 γ-ray events were accompanied by type III bursts, and 23/25 by type II bursts. The type III bursts show that electrons (particles) had access to open field lines during all γ-ray events. The bursts occurred in general during the impulsive and early post-impulsive phase of the parent flare, as commonly observed in SEP events. They were not found later on during long-duration γ-ray events. The presence of type II bursts shows that shock waves in the high corona are a common counterpart of γ-ray events. Metric continuum emission (type IV bursts) is observed in some events.

8.4.1 Impulsive and Early Post-impulsive γ-Ray Emission

A rare *Fermi*/LAT observation showing a γ-ray event during and after the impulsive flare phase is illustrated in Fig. 8.4. Hard X-rays (INTEGRAL/ACS, photon energies >80 keV; Rodríguez-Gasén et al. 2014) rise and decline during the rise of the soft X-ray flux, as do the radio flux densities at 15,400 and 8800 MHz. The γ-ray emission, which is observed with much poorer time coverage, shows a similar rise and initial decay, but then stays on an enhanced level until the end of

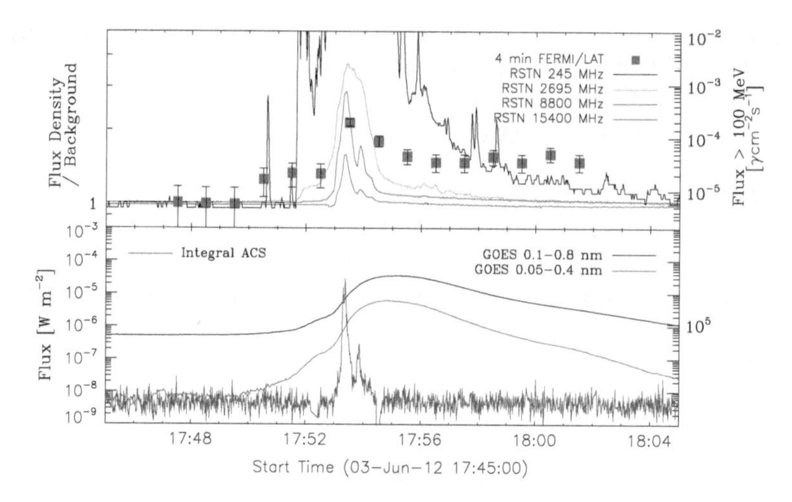

Fig. 8.4 Time evolution of the high-energy γ-ray, X-ray and radio emission in the impulsive and early post-impulsive phase of a flare

the observations. This long persistence is the key feature of the sustained events introduced in Sect. 8.3. While the microwaves $\geq$8800 MHz show no counterpart, emissions at 2695 and 245 MHz have a prolonged tail that accompanies the observed part of the post-impulsive γ-ray emission. Renewed energy release after the impulsive phase is also suggested by the bump in the decaying soft X-ray profile. The continued emission of both pion-decay γ-rays and microwaves in the early post-impulsive phase is consistent with the earlier observations by GAMMA1 (Kocharov et al. 1994; Akimov et al. 1996) and SMM (Trottet et al. 1994).

8.4.2 Long-Duration γ-Ray Events

As shown in Fig. 8.2, γ-ray emission may rise again well after the impulsive phase. Among the 25 γ-ray events studied, a total of twelve had duration longer than 2 h. Long-duration γ-ray emission occurs most often during the decay of the associated soft X-ray burst, as illustrated in Fig. 8.5. In EUV images taken by AIA/SDO or SWAP/Proba2 the decay of the soft X-ray bursts is accompanied by the formation of post-flare loop arcades.

Although there is still some microwave burst activity at low frequencies (2.695 MHz) in Fig. 8.5, the long-duration γ-rays are in general not accompanied by time-extended or recurrent hard X-ray or microwave burst activity. On 2011 Mar 7 the long-duration γ-rays are accompanied by several flares, rather than the decay of a single soft X-ray burst. But the microwave or hard X-ray emissions are weak, and in any case are only observed during a short fraction of the γ-ray event. In

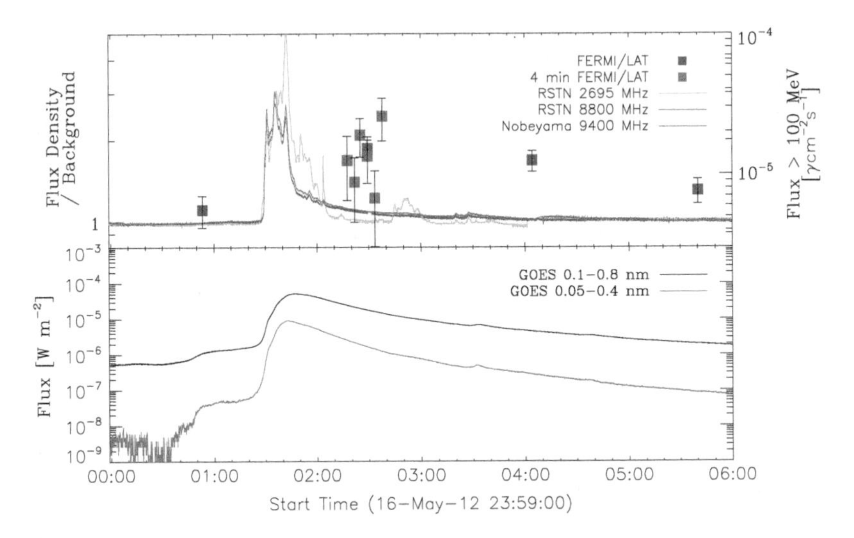

Fig. 8.5 Time evolution of the high-energy γ-ray, soft and hard X-ray and radio emission from the impulsive phase throughout the rise and decay of sustained γ-ray emission

summary, we find no evidence of repeated efficient electron acceleration in the low corona during most parts of this or any other long-duration γ-ray event.

8.4.3 *Soft X-Ray Bursts and γ-Ray Events*

The close relationship between the durations of γ-ray and soft X-ray emissions is confirmed by the quantitative analysis of the decay profiles of soft X-ray bursts associated with γ-ray events: short duration γ-ray bursts (<2 h; the exact duration is difficult to determine because of gaps in the solar observations of *Fermi*/LAT) are found to have relatively short soft X-ray decay phases, while the long γ-ray events accompany comparatively long soft X-ray decay phases. There are some intermediate events, showing the distribution of burst durations is not bimodal. But it is clear from this observational analysis that long-duration γ-ray events tend to be associated with long-duration energy release in the solar corona.

Although the long-duration γ-ray emission may start more than an hour after the peak of the associated soft X-ray burst, the logarithms of the peak fluxes of the two emissions correlate, with a linear correlation coefficient 0.72, and a probability that the same or a higher value is obtained from two unrelated samples of 0.8%. The corresponding values for the CME speed and the logarithm of the γ-ray peak flux are 0.61 and 3.7%, respectively. These are rough evaluations, because the γ-ray time profile is not densely covered by the *Fermi*/LAT observations, and the peak fluxes are only lower limits.

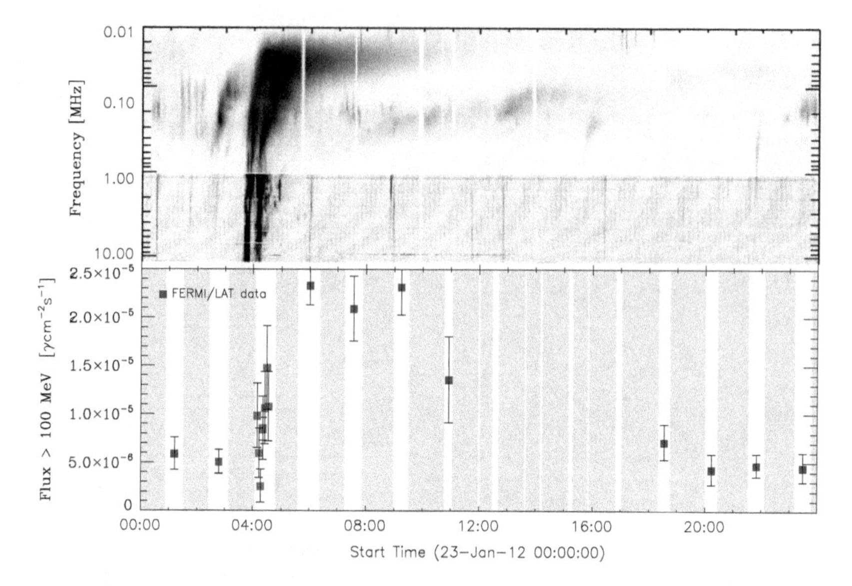

Fig. 8.6 Time history of high-energy γ-rays and the decametric-hectometric radio emission, showing a type II burst accompanies the γ-rays. The *grey vertical stripes* in the *bottom panel* show time intervals of unfavourable solar observing conditions for *Fermi*/LAT

8.4.4 Coronal Shock Waves and γ-Ray Events

The only signature of electron acceleration that we found to accompany long-lasting γ-ray emission are DH type II bursts, which are ascribed to CME-driven shocks (Reiner et al. 2007). The type II burst in Fig. 8.6 mostly shows up as a sequence of patches after ~7 UT that gradually drift from higher to lower frequencies. The noticeable fact is that this emission persists throughout the several hours duration of the γ-ray emission.

It is tempting to interpret the association between sustained γ-ray events and DH type II bursts as evidence that the CME shock accelerates protons to relativistic energies, which then stream back from the downstream region of the shock to the low solar atmosphere, where they create pions (see Sect. 8.3 and Chap. 9). We consider in the following the implication of this interpretation on the number of particles that must be present in the downstream region of the shock in order to account for the observed γ-ray emission. To this end we analyze the events with a long duration sustained emission that occurred on 2011 Mar 7, 2012 Jan 23, Mar 5 and May 17.

All four events were accompanied by CMEs well-observed by the LASCO coronagraphs. The quasi-parallel shock often invoked as the site of SEP acceleration is expected to be located around the summit of the CME, following the trajectory shown by the height-time curves given in the LASCO CME catalogue.[1] The type II source is usually believed to be located in the quasi-perpendicular region of the shock (Pulupa and Bale 2008), hence on the flank of the CME rather than near its summit. We therefore considered separately the heights of the CME estimated by the LASCO catalogue and of the type II source, inferred from the electron density model of Leblanc et al. (1999), both at the peak of the long-duration γ-ray emission. The heliocentric distances of the CME apices range from 11 $R_\odot$ to 26 $R_\odot$, those of type II sources from 7 $R_\odot$ to 18 $R_\odot$.

Particles accelerated at the shock must hence stream to the solar chromosphere over a distance of many solar radii without being mirrored by the magnetic field before. This implies that they are injected into coronal magnetic field lines near the shock with a pitch angle smaller than some limiting value α_0 such that $\sin^2\alpha_0 = B(r)/B(R_\odot)$. The magnetic field decreases outward from the low corona with some power of the heliocentric distance r. If we just extrapolate the radial component of the solar wind magnetic field back to the photosphere, hence assuming that $B(r) \sim r^{-2}$, the maximum pitch angles allowed for protons that reach the low solar atmosphere are 2°–5° when the particles start from the CME apex, and 3°–8° when they start from the type II source. Hence only protons injected nearly parallel to the coronal magnetic field lines near the shock can reach a sufficiently dense part of the solar atmosphere to undergo nuclear reactions. If we suppose that the protons are isotropically distributed at the shock, only a fraction $1-\cos\alpha_0$ is expected to achieve this. This fraction is 1% of the initial population for $\alpha_0 = 8°$ ($6 \cdot 10^{-4}$ for $\alpha_0 = 2°$). It will be further reduced by the expected stronger decrease of the magnetic field within the solar wind source surface.

8.5 Solar Energetic Particle Events Associated with *Fermi*/LAT Gamma-Ray Events

All *Fermi*/LAT γ-ray events discussed in Sect. 8.3 are associated with SEP events observed at 1 AU from the Sun, by instruments at L1 or by the twin STEREO spacecraft, or both. Proton and/or electron enhancements are observed with SoHO/ERNE (Torsti et al. 1995), STEREO/LET (Mewaldt et al. 2008), STEREO/HET (von Rosenvinge et al. 2008), STEREO/SEPT (Müller-Mellin et al. 2008), and ACE/EPAM (Gold et al. 1998) over a wide range of energy (55 keV to 4 MeV for electrons and 1.6 to 130 MeV for protons). Higher energy protons are available from SoHO/EPHIN (penetrating protons at 100–1000 MeV energies) as well as from GOES/HEPAD (only for 9 of the 25 investigated *Fermi*/LAT events, see Fig. 8.7 for an example) in three differential energy channels ranging from 330 to 700 MeV and an integral channel >700 MeV.

[1] https://cdaw.gsfc.nasa.gov/CME_list/.

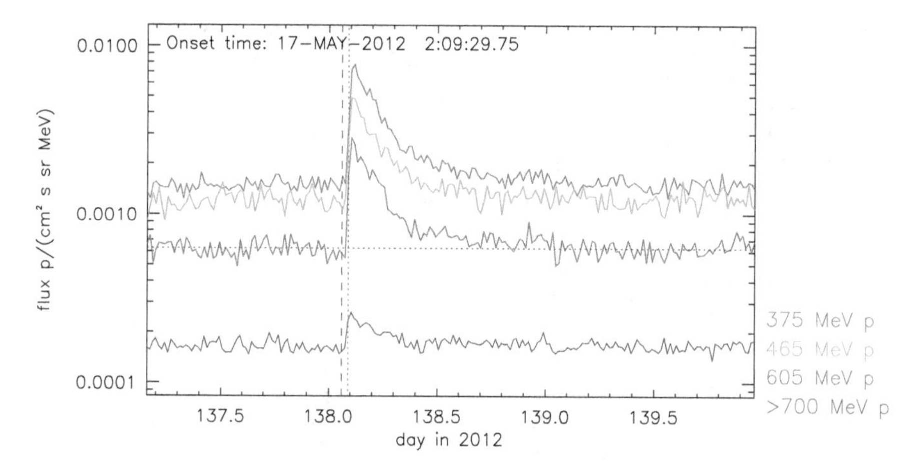

Fig. 8.7 GOES 15/HEPAD proton fluxes on 2012 May 17. The *vertical purple dashed* and the *red dotted lines* indicate respectively the flare onset and the derived SEP onset with the sigma-method (see text)

8.5.1 SEP Characteristics and Association with Fermi/LAT

Most of the SEP events show a slow rise phase, but some show a fast rise phase in all observed energy ranges (see Reames 1999 for a review on general SEP characteristics). This distinction is somewhat subjective, but we mean by "fast" a rise in 2–3 h, and by "slow" a rise within several hours, sometimes one day. Onset times for all observed SEP events can be derived either by implementing a threshold to be exceeded by the flux as described in Malandraki et al. (2012) or a Poisson-CUSUM analysis (Huttunen-Heikinmaa et al. 2005). For SEP events with a fast rise a velocity dispersion analysis (VDA) can be used (Krucker et al. 1999; Malandraki et al. 2012) to estimate both the release time at the Sun and the apparent path length of the particles: the solar release time of the first particles seen at the spacecraft is obtained from a linear fit of the derived onset times of electrons and protons at different energies as a function of their respective inverse velocity. Such an analysis could only be performed for 11 of the 25 *Fermi*/LAT events. For a comparison of derived release times with photon arrival times measured at the observer's distance (1 AU) we add 8 min.

Hereafter, we will discuss the events of 2011 Mar 7, 2012 Mar 7, 2013 Apr 11 and 2014 Feb 25, for which a VDA analysis could be performed at L1 and/or STEREO, as well as the events of 2012 Jan 23 and May 17 that are of particular interest for numerical transport simulations discussed in Chap. 9. Table 8.1 gives the date and location, soft X-ray class, start and peak time, start and end of the decametric-to-hectometric type III bursts and CME speed from the LASCO/CME catalogue. SEP parameters are given for STEREO A and B (STA, STB), for SoHO/ERNE or ACE/EPAM at L1, and for GOES/HEPAD (HEP) such as the characterization of the

Table 8.1 Characteristics of the 6 SEP event cases discussed in this section (see text for further information)

Date	Soft X-rays Class Location	Start Peak	DH III	CME [km s^{-1}]	SEP (s/c, rise, energy)	SRT + 8 min [min]
2011 Mar 7	M3.7	19:43	19:50–20:10	2125	STA (s) ~60 MeV	–
	N30W47	20:58			L1 (f) ~ 80 MeV	20:26±00:04
					HEP no detection	–
					STB (f) ~60 MeV	–
23 Jan 2012	M8.7	03:38	03:40–04:20	2175	STA (s) ~100 MeV	–
	N33W21	04:34			L1 (f) ~100 MeV	–
					HEP (s) ~ 600 MeV weak	–
					STB (s) ~100 MeV	–
2012 Mar 7	X5.4	00:02	00:15–01:00	2864	STA (s) > 130 MeV	–
	N17E27	00:40			L1 (s) >100 MeV	23:21±00:51
	X1.2	01:05	01:15–01:30	1825	HEP (s) >700 MeV	–
	N17E27	01:23			STB (f) >100 MeV	00:37±00:01
2012 May 17	M5.1	01:25	01:30–01:40	1582	STA (s) ~100 MeV	–
	N07W88	02:14			L1 (f) ~ 130 MeV	–
					HEP (f) >700 MeV	–
					STB (s) ~100 MeV	–
2013 Apr 11	M6.5	06:55	07:00–07:30	861	STA (s) weak	–
	N07E13	07:29			L1 (f) ~ 130 MeV	07:24±00:08
					HEP very weak ~600 MeV	–
					STB (f) ~100 MeV	07:10±00:01
25 Feb 2014	X4.9	00:39	00:45–01:15	2147	STA (f) ~100 MeV	01:06±00:01
	N00E78	01:03			L1 (s) ~ 130 MeV	01:22±00:20
					HEP very weak	–
					STB (f) ~100 MeV	00:56±00:01

SEP flux rise as "fast" (f) or "slow" (s), the approximate energy up to which the SEPs are observed and the derived solar release time (SRT) from VDA.

2011 Mar 7: The event had moderate SEP flux. The *Fermi*/LAT sustained emission (Fig. 8.2) was observed ~15 min after the CME onset (defined from inspection of SDO/AIA images) and peaked several hours later (~7 h). The derived SRT from the L1 observations is ~20 min later than the *Fermi*/LAT emission onset and the particle path length of ~1.9 AU does not indicate substantial scattering.

2012 Jan 23: An intense SEP event was observed by ACE/EPAM at L1. VDA could not produce reliable SRT results, while GOES/HEPAD indicates a very weak event with a poorly determined onset (~7 h after the flare). The weak event was preceded by a slow rise right after the flare. The interaction between the parent CME and a preceding much slower one (1400 km s^{-1}), associated with an earlier

M1.1 class flare, may be responsible for both the observed high SEP flux and the significant changes in the SEP intensity profile (Joshi et al. 2013). Sustained *Fermi*/LAT emission was observed ~15 min after the hard X-ray flare peak, lasting for many hours while the delayed observed GOES/HEPAD onset with respect to the *Fermi*/LAT onset could be related to this several-hours lasting sustained γ-ray emission.

2012 Mar 7: For the double solar event with considerable SEP flux, GOES/HEPAD saw a steep rise ~6 h after the first flare. It was, however, preceded by a continuous slow rise starting immediately after the flare, as often observed in eastern SEP events. According to Kouloumvakos et al. (2016) (a) the first flare/CME was responsible for the SEP event observed at different spacecraft, (b) the proton SRT observed by STEREO B is consistent with the arrival of an observed EUV wave on the Sun at the STEREO-B footpoint, and (c) the considerably delayed SRT at L1 compared to STEREO B suggests a release of particles further away from the Sun consistent with the timing and location of the shock's western flank. No plausible explanations could be derived from the observed behaviour at STEREO A. Our L1 VDA analysis is not reliable as the derived apparent particle path length is ~8 AU. The observed large *Fermi*/LAT sustained fluxes, peaking after the first flare and lasting for several hours, have a derived first onset near the first flare peak time, while the second *Fermi*/LAT onset is estimated ~45 min after the second flare onset. The derived STEREO B release time is co-temporal, within errors, with the first *Fermi*/LAT onset.

2012 May 17: The strong SEP event was associated with the first Ground Level Enhancement (GLE) of solar cycle 24. VDA could not produce reliable SRTs. GOES/HEPAD registered a fast rising intense event starting right after the flare (see Fig. 8.7). The expansion of the shock forming in the corona was studied using a new technique based on coronal magnetic field reconstructions, full magneto-hydrodynamic simulations and multi-point imaging inversion techniques (Rouillard et al. 2016). This analysis concluded that GeV particles were released when this shock became super-critical (Mach numbers >3). The magnetic connectivity between the shock and L1 was established via a magnetic cloud that erupted from the same active region 5 days earlier. *Fermi*/LAT observations (Fig. 8.5) indicate significant sustained emission starting after the flare and lasting several hours, while the GOES/HEPAD flux onset coincides with the *Fermi*/LAT emission onset.

2013 Apr 11: A fast-rising strong SEP event was observed at L1 and STEREO B, but only a small slowly rising one at STEREO A. GOES/HEPAD recorded a very weak event with a poorly-determined onset 3 h after the flare. By determining the angular extent of the observed EUV wave and CME, Lario et al. (2014) concluded that while the particle SRT from STEREO B is within uncertainties consistent with the arrival of the EUV wave and CME-driven shock at the footpoint of the spacecraft, the EUV wave did not reach the footpoint of the field lines connecting to L1; the observed intense SEPs at L1 were most likely originating from the western flank of the CME-driven shock as it was propagating higher in the corona. Our analysis indicates (a) a particle release time to L1 that coincides within errors with the peak of the flare and is at least 15 min later than the SRT from STEREO B and

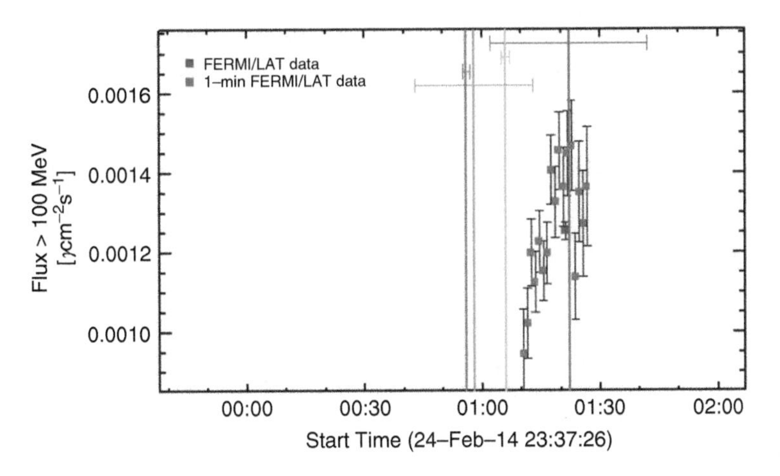

Fig. 8.8 *Fermi*/LAT observations for the 2014 Feb 25 event. *Vertical light and dark green lines* indicate respectively the derived STEREO A and B onsets, while *purple and orange vertical lines* indicate the respective L1 and *Fermi*/LAT onsets. *Horizontal lines* indicate the derived error ranges

(b) an apparent path length of ~2.1 AU, consistent with the picture emerging from Lario et al. (2014). The derived STEREO B SRT coincides precisely with the onset of the *Fermi*/LAT sustained emission that lasts for ~20 min and peaks a few minutes after the soft X-ray flux.

25 Feb 2014 (Fig. 8.8): The SEPs rose fast to high fluxes at STEREO A and B, while a weak slowly-rising event was recorded at L1. GOES/HEPAD showed a very weak SEP event with a poorly-determined onset >5 h after the flare. According to Lario et al. (2016), despite the considerable distance between the footpoints of the field lines connecting the Sun with STEREO A, B and L1 and the small extent of the observed EUV wave, the expansion of the extended shock, accompanying the CME, to higher latitudes into the corona determined the release of particles and the observed intense SEP event. Our analysis indicates a SRT at both STEREO satellites and L1 well after the flare in accordance to these findings. However, the VDA at L1 and the respective SRT are not reliable as the particle path length is ~5.2 AU. Due to gaps in the observing time, sustained γ-ray emission is only observed for a short time, starting ~15 min after the onset of the flare, then peaking and dropping quite rapidly within ~20 min. The STEREO B SRT coincides with the *Fermi* onset, while the STEREO A SRT is 10 min later.

8.5.2 SEP Spectra

Event-integrated energy spectra of oxygen, neon, and iron+nickel as a group, and the abundance ratios Ne/O, (Fe+Ni)/O, and (Fe+Ni)/Ne were investigated by using

SOHO/ERNE measurements. Heavy ion signatures were searched for in all 25 γ-ray events, and were observed above the quiet-time background in twelve of them. For events occurring in a close sequence of time it was not possible to distinguish possible multiple injections of heavy ions that could have been associated with each individual γ-ray event. This was the case, e.g., for the four events during the time period of 2012 Mar 7–10. The heavy ions observed were associated with the first γ-ray event of this period. Continuous data coverage over an entire event was also required, and this limited the investigation to eight of the 25 γ-ray events.

The event-integrated intensities of O, Ne and Fe+Ni were calculated for 10 energy channels between ~3.2–160 MeV/nucleon of each ion species by integrating from the event onset in each energy channel till the time when the background-detrended cumulative intensity at that energy reached 95% of the maximum. Thus, individual integration times were used for each ion at each energy channel. The energy spectra were fitted by using double power-law functions by Band et al. (1993). The fitting procedure gave the low and high-energy spectral indices (γ_a and γ_b), the break energy (E_B), and the normalization constant for the spectra. These quantities for oxygen and iron+nickel are given in Table 8.2 for the events of 2012 Mar 7 and 2013 Apr 11. The measured abundance ratios Ne/O and (Fe+Ni)/O as function of energy are shown in Fig. 8.9. The γ-ray events were characterized by the peak fluxes of $\sim 94 \times 10^{-5}$ cm^{-2} s^{-1} (2012 Mar 7) and $\sim 3.9 \times 10^{-5}$ cm^{-2} s^{-1} (2013 Apr 11) with event durations of 9.9 h and 1.2 h, respectively. Both were eastern events (Table 8.1).

Figure 8.9 shows that the Ne/O ratios in these two events are similar and roughly constant (~0.2) up to ~30 MeV/nucleon, but differ at higher energies with the ratio decreasing with energy in the 2013 Apr 11 event. On the other hand, the (Fe+Ni)/O ratios are quite strongly increasing with energy. The ratio is significantly higher in the event of 2013 Apr 11 than on 2012 Mar 7 reaching the maximum of ~0.7 in the 8–20 MeV/nucleon energy range and then decreasing with energy. The changes with energy in the abundance ratios and the differences between the two studied events can be explained by the differences in the energy where the high-energy spectral index is predominant. The high-energy spectral indices are believed to be dependent on the contribution of suprathermal flare material in the seed population. The high (Fe+Ni)/O ratio in the 2013 Apr 11 event indicates a more impulsive nature and larger flare material contribution in this smaller γ-ray event. It should be noted, however, that both of these events were large proton events lasting for several days. The abundance ratios can be compared with the average values of Ne/O = 0.157 and Fe/O = 0.131 in gradual solar energetic particle events at energies 2–15

Table 8.2 Heavy ion spectral parameters for the events of 2012 Mar 7 and 2013 Apr 11

	Oxygen			Iron+Nickel		
Event	γ_a	γ_b	E_B (MeV)	γ_a	γ_b	E_B (MeV)
2012 Mar 7	1.03	5.22	9.8	0.3	4.9	6.4
2013 Apr 11	1.53	4.01	18	−0.2	3.1	3

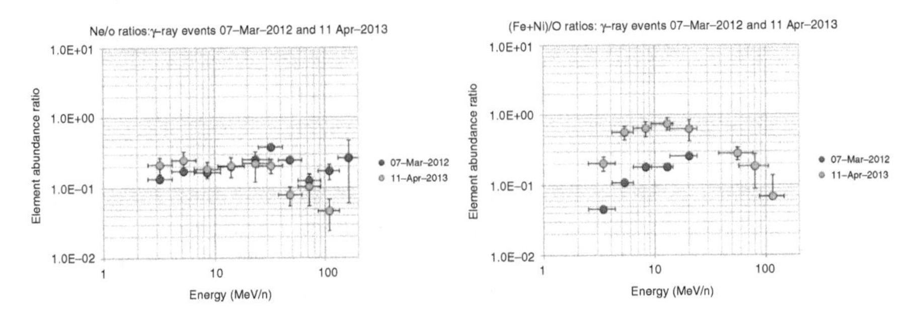

Fig. 8.9 Abundance ratios Ne/O and (Fe+Ni)/O for the events of 2012 Mar 7 and 2013 Apr 11

MeV/nucleon (Reames 2014) and Ne/O = 0.478 and Fe/O = 1.17 on the average in impulsive SEP events (Reames et al. 2014).

8.6 Summary and Discussion

This overview on electromagnetic and SEP signatures that accompany sustained γ-ray emission of 25 events detected by *Fermi*/LAT is summarized as follows:

- High-energy γ-ray emission during the impulsive and early post-impulsive phase is accompanied by hard X-ray and radio signatures of energetic electron acceleration in the solar atmosphere. This is similar to the earlier findings for pion-decay γ-ray events (Kocharov et al. 1994; Trottet et al. 1994; Akimov et al. 1996) and GLEs (e.g., Klein et al. 2014). The manifestations in the early post-impulsive phase are visible at lower microwave frequencies than those of the impulsive phase, and at dm-to-m wavelengths (type IV bursts).
- γ-ray emission lasting several hours is not accompanied by similarly extended emissions in hard X-rays or radio waves. In all cases (but 2011 Mar 7) these long-duration events accompany the gradual decay of the soft X-ray flux. Although the parent soft X-ray burst starts sometimes more than an hour before the long-lasting γ-ray enhancement, there are statistical relationships between (1) the importance of the parent soft X-ray burst and the peak flux of the long-duration γ-ray emission, and (2) the duration of the decay phase of the soft X-ray burst and the duration of the long-duration γ-ray emission.
- The 2011 Mar 7 γ-ray event was accompanied by several distinct flares. But only the first of them showed conspicuous hard X-ray and radio emission. There is no indication that the long-lasting γ-ray emission could be understood as the superposition of several successive acceleration episodes in independent flares.
- Besides decaying soft X-ray emission and the related post-flare loop arcades, the only other electromagnetic counterpart of long-duration γ-rays are decametric-to-hectometric type II bursts. They accompany the γ-ray emission of all long-duration events.

- SEP events are associated with all high-energy γ-ray events. In most well-connected γ-ray events with long duration GOES/HEPAD detected protons above 300 MeV, and in several cases above 700 MeV. The 2011 Mar 7 γ-ray event is again an exception.
- At energies of several MeV to about 100 MeV the SEP events are strong, but we found no characteristic feature that distinguishes them from strong SEP events in general. All are associated with DH type III bursts, as are SEP events in general (Cane et al. 2010; Vainio et al. 2013). When the solar release time of the first SEPs could be inferred from velocity dispersion analysis, it was found close to the interval of the type III bursts. A temporal relationship with the rise of the late long-duration γ-ray emission could not be excluded, but definite evidence for such a relationship could not be established either.
- The event-integrated abundances of O, Ne, Fe+Ni, and ^{3}He in the analysed SEP events show no unambiguous classification with respect to the conventional impulsive or gradual SEP values. There are, however, significant event-to-event variations in the ^{3}He and heavy ion abundance ratios and in their energy dependence. The sources of these variations and their possible relationships to the characteristics of the γ-ray emission need further analysis.

We conclude from the observations that pion-decay γ-rays during the impulsive and early post-impulsive phase are a manifestation of common particle acceleration with electrons, seen through hard X-ray and radio emissions. In the post-impulsive phase the signatures are mostly seen at long-centimetric to metric waves.

The association of γ-ray events with soft X-ray emission from the thermal plasma is not a direct clue to the origin of the mildly relativistic protons. Particle acceleration at the CME-driven shock in the high corona appears as an attractive interpretation, because it can explain that γ-rays are observed from flares behind the solar limb (Sect. 8.3) and because type II bursts are the unique systematic radio counterpart of long-duration γ-ray emissions (Sect. 8.4). Stereoscopic CME observations and sophisticated modelling allow one to estimate Mach numbers. Modelling of the shock acceleration (Chap. 9) shows that the highest Mach numbers, observed in restricted regions on the CME surface (Rouillard et al. 2016), may explain shock-accelerated protons up to GeV energies. This modelling shows furthermore that magnetic connections exist between the CME shock and the solar surface (Plotnikov et al. 2017). The interpretation faces two challenges from our present analysis: First, only a small fraction of the particles at the shock, less than 1% as estimated by a simple model, can reach the low solar atmosphere where nuclear interactions can take place. This may still be consistent with the estimation of proton numbers in space and the low solar atmosphere mentioned in Sect. 8.3. The second challenge is to explain why SEPs were only observed up to about 80 MeV during the long-duration γ-ray event on 2011 Mar 7.

The HESPERIA sample of 25 events is too small to draw firm general conclusions on the origin of high-energy events and their relationship with SEPs. But it is rich enough to provide constraints on the processes of acceleration and coronal transport of mildly relativistic protons.

Acknowledgements The authors are grateful to Gerald Share and Ron Murphy for preparing the text of Sect. 8.3.

References

Abdo, A.A., Ackermann, M., Ajello, M., et al.: Astrophys. J. **734**, 116 (2011). doi:10.1088/0004-637X/734/2/116
Ackermann, M., Ajello, M., Allafort, A., et al.: Astrophys. J. **745**, 144 (2012a). doi:10.1088/0004-637X/745/2/144
Ackermann, M., Ajello, M., Allafort, A., et al.: Astrophys. J. **748**, 151 (2012b). doi:10.1088/0004-637X/748/2/151
Ackermann, M., Ajello, M., Albert, A., et al.: Astrophys. J. **787**, 15 (2014). doi:10.1088/0004-637X/787/1/15
Ackermann, M., Allafort, A., Baldini, L., et al.: Astrophys. J. **835**, 219 (2017). doi:10.3847/1538-4357/835/2/219
Ajello, M., Albert, A., Allafort, A., et al.: Astrophys. J. **789**, 20 (2014). doi:10.1088/0004-637X/789/1/20
Akimov, V.V., Afanasyev, V.G., Belaousov, A.S., et al.: Sov. Astron. Lett. **18**, 69 (1992)
Akimov, V.V., Ambrož, P., Belov, A.V., et al.: Sol. Phys. **166**, 107 (1996)
Alexander, D., Dunphy, P.P., MacKinnon, A.L.: Sol. Phys. **151**, 147 (1994). doi:10.1007/BF00654088
Atwood, W.B., Abdo, A.A., Ackermann, M., et al.: Astrophys. J. **697**, 1071 (2009). doi:10.1088/0004-637X/697/2/1071
Band, D., Matteson, J., Ford, L., et al.: Astrophys. J. **413**, 281 (1993). doi:10.1086/172995
Bougeret, J.L., Kaiser, M.L., Kellogg, P.J., et al.: Space Sci. Rev. **71**, 231 (1995). doi:10.1007/BF00751331
Bougeret, J.L., Goetz, K., Kaiser, M.L., et al.: Space Sci. Rev. **136**, 487 (2008). doi:10.1007/s11214-007-9298-8
Cane, H.V., Richardson, I.G., von Rosenvinge, T.T.: J. Geophys. Res. (Space Phys.) **115**, A08101 (2010). doi:10.1029/2009JA014848
Chupp, E.L., Ryan, J.M.: Res. Astron. **9**, 11 (2009). doi:10.1088/1674-4527/9/1/003
Debrunner, H., Lockwood, J.A., Barat, C., et al.: Astrophys. J. **479**, 997 (1997)
Dunphy, P.P., Chupp, E.L., Bertsch, D.L., et al.: Sol. Phys. **187**, 45 (1999)
Forrest, D.J., Vestrand, W.T., Chupp, E.L., et al.: Int. Cosmic Ray Conf. **4**, 146 (1985)
Forrest, D.J., Vestrand, W.T., Chupp, E.L., et al.: Adv. Space Res. **6**(6), 115 (1986). doi:10.1016/0273-1177(86)90127-4
Gold, R.E., Krimigis, S.M., Hawkins, S.E., et al.: Space Sci. Rev. **86**, 541 (1998). doi:10.1023/A:1005088115759
Huttunen-Heikinmaa, K., Valtonen, E., Laitinen, T.: Astron. Astrophys. **442**, 673 (2005). doi:10.1051/0004-6361:20042620
Joshi, N.C., Uddin, W., Srivastava, A.K., et al.: **52**, 1 (2013). doi:10.1016/j.asr.2013.03.009
Kanbach, G., Bertsch, D.L., Fichtel, C.E., et al.: Astron. Astrophys. Suppl. **97**, 349 (1993)
Klein, K.L., Masson, S., Bouratzis, C., et al.: Astron. Astrophys. **572**, A4 (2014). doi:10.1051/0004-6361/201423783
Kocharov, L.G., Kovaltsov, G.A., Kocharov, G.E., et al.: Sol. Phys. **150**, 267 (1994)
Kocharov, L., Debrunner, H., Kovaltsov, G., et al.: Astron. Astrophys. **340**, 257 (1998)
Kocharov, L., Laitinen, T., Vainio, R., et al.: Astrophys. J. **806**, 80 (2015). doi:10.1088/0004-637X/806/1/80
Kouloumvakos, A., Patsourakos, S., Nindos, A., et al.: Astrophys. J. **821**, 31 (2016). doi:10.3847/0004-637X/821/1/31
Krucker, S., Larson, D.E., Lin, R.P., Thompson, B.J.: Astrophys. J. **519**, 864 (1999)

Kuznetsov, S.N., Kurt, V.G., Yushkov, B.Y., et al.: Sol. Phys. **268**, 175 (2011). doi:10.1007/s11207-010-9669-2

Kuznetsov, S.N., Kurt, V.G., Yushkov, B.Y., et al.: In: Kuznetsov V. (ed.) The Coronas-F Space Mission. Astrophysics and Space Science Library, vol. 400, p. 301. Springer, Berlin (2014). doi:10.1007/978-3-642-39268-9_10

Lario, D., Raouafi, N.E., Kwon, R.Y., et al.: Astrophys. J. **797**, 8 (2014). doi:10.1088/0004-637X/797/1/8

Lario, D., Kwon, R.Y., Vourlidas, A., et al.: Astrophys. J. **819**, 72 (2016). doi:10.3847/0004-637X/819/1/72

Leblanc, Y., Dulk, G.A., Bougeret, J.L.: In: Habbal, S.R., Esser, R., Hollweg, J.V., Isenberg, P.A. (eds.) Solar Wind Nine. American Institute of Physics Conference Series, vol. 471, pp. 83–86. AIP Press, New York (1999). doi:10.1063/1.58787

Lin, R.P., Dennis, B.R., Hurford, G.J., et al.: Sol. Phys. **210**, 3 (2002). doi:10.1023/A:1022428818870

Lockwood, J.A., Debrunner, H., Ryan, J.M.: Sol. Phys. **173**, 151 (1997). doi:10.1023/A:1004908209975

Malandraki, O.E., Agueda, N., Papaioannou, A., et al.: Sol. Phys. **281**, 333 (2012). doi:10.1007/s11207-012-0164-9

Mandzhavidze, N., Ramaty, R.: Astrophys. J. Lett. **396**, L111 (1992). doi:10.1086/186529

Mewaldt, R.A., Cohen, C.M.S., Cook, W.R., et al.: Space Sci. Rev. **136**, 285 (2008). doi:10.1007/s11214-007-9288-x

Müller-Mellin, R., Böttcher, S., Falenski, J., et al.: Space Sci. Rev. **136**, 363 (2008). doi:10.1007/s11214-007-9204-4

Murphy, R.J., Dermer, C.D., Ramaty, R.: Astrophys. J. Supp. Ser. **63**, 721 (1987). doi:10.1086/191180

Murphy, R.J., Share, G.H., Grove, J.E., et al.: Astrophys. J. **490**, 883 (1997). doi:10.1086/304902

Nakajima, H., Sekiguchi, H., Sawa, M., et al.: Publ. Astron. Soc. Jpn. **37**, 163 (1985)

Pesce-Rollins, M., Omodei, N., Petrosian, V., et al.: Astrophys. J. Lett. **805**, L15 (2015). doi:10.1088/2041-8205/805/2/L15

Plotnikov, I., Rouillard, A.P., Share, G.H.: *Astron. Astrophys.* in press, (2017). doi:10.1051/0004-6361/201730804

Pulupa, M., Bale, S.D.: Astrophys. J. **676**, 1330 (2008). doi:10.1086/526405

Rank, G., Ryan, J., Debrunner, H., et al.: *Astron. Astrophys.* **378**, 1046 (2001). doi:10.1051/0004-6361:20011060

Reames, D.V.: Space Sci. Rev. **90**, 413 (1999)

Reames, D.V.: Sol. Phys. **289**, 977 (2014). doi:10.1007/s11207-013-0350-4

Reames, D.V., Cliver, E.W., Kahler, S.W.: Sol. Phys. **289**, 3817 (2014). doi:10.1007/s11207-014-0547-1

Reiner, M.J., Kaiser, M.L., Bougeret, J.L.: Astrophys. J. **663**, 1369 (2007). doi:10.1086/ 510827

Rodríguez-Gasén, R., Kiener, J., Tatischeff, V., et al.: Sol. Phys. **289**, 1625 (2014). doi:10.1007/s11207-013-0418-1

von Rosenvinge, T.T., Reames, D.V., Baker, R., et al.: Space Sci. Rev. **136**, 391 (2008). doi:10.1007/s11214-007-9300-5

Rouillard, A.P., Plotnikov, I., Pinto, R.F., et al.: Astrophys. J. **833**, 45 (2016). doi:10.3847/1538-4357/833/1/45

Ryan, J.M.: Space Sci. Rev. **93**, 581 (2000)

Ryan, J.M., Lee, M.A.: Astrophys. J. **368**, 316 (1991). doi:10.1086/169695

Ryan, J., Forrest, D., Lockwood, J., et al.: In: Ryan, J., Vestrand, W.T. (eds.) High-Energy Solar Phenomena - a New Era of Spacecraft Measurements. American Institute of Physics Conference Series, vol. 294, pp. 89–93 (1994). doi:10.1063/1.45205

Schlemm, C.E., Starr, R.D., Ho, G.C., et al.: Space Sci. Rev. **131**, 393 (2007). doi:10.1007/s11214-007-9248-5

Share, G.H., Murphy, R.J., Tolbert, K., et al.: Astrophys. J. (2017, Submitted)

Torsti, J., Valtonen, E., Lumme, M., et al.: Sol. Phys. **162**, 505 (1995). doi:10.1007/BF00733438

Torii, C., Tsukiji, Y., Kobayashi, S., et al.: Proceedings of the Research Institute of Atmospherics, Nagoya University, vol. 26, p. 129 (1979)
Trottet, G., Chupp, E.L., Marschhaeuser, H., Pick, M., Soru-Escaut, I., Rieger, E., Dunphy, P.P.: Astron. Astrophys. **288**, 647 (1994)
Vainio, R., Valtonen, E., Heber, B., et al.: J. Space Weather Space Clim. **3**, A12 (2013). doi:10.1051/swsc/2013030
Vilmer, N., MacKinnon, A.L., Trottet, G., Barat, C.: Astron. Astrophys. **412**, 865 (2003). doi:10.1051/0004-6361:20031488
Vilmer, N., MacKinnon, A.L., Hurford, G.J.: Space Sci. Rev. **159**, 167 (2011). doi:10.1007/s11214-010-9728-x

Chapter 9
Modelling of Shock-Accelerated Gamma-Ray Events

Alexandr Afanasiev, Angels Aran, Rami Vainio, Alexis Rouillard, Pietro Zucca, David Lario, Suvi Barcewicz, Robert Siipola, Jens Pomoell, Blai Sanahuja, and Olga E. Malandraki

Abstract Solar γ-ray events recently detected by the *Fermi*/LAT instrument at energies above 100 MeV have presented a puzzle for solar physicists as many of such events were observed lasting for many hours after the associated flare/coronal mass ejection (CME) eruption. Data analyses suggest the γ-ray emission originate from decay of pions produced mainly by interactions of high-energy protons deep in the chromosphere. Whether those protons are accelerated in the associated flare or in the CME-driven shock has been under active discussion. In this chapter, we present some modelling efforts aimed at testing the shock acceleration hypothesis.

A. Afanasiev (✉) • R. Vainio • S. Barcewicz • R. Siipola
Department of Physics and Astronomy, University of Turku, 20014 Turku, Finland
e-mail: alexandr.afanasiev@utu.fi; rami.vainio@utu.fi; t09susaa@utu.fi; roamsi@utu.fi

A. Aran • B. Sanahuja
Departament de Física Quàntica i Astrofísica, Institut de Ciències del Cosmos (ICCUB), Universitat de Barcelona, Barcelona, Spain
e-mail: angels.aran@fqa.ub.edu; blai.sanahuja@ub.edu

A. Rouillard
Institut de Recherche en Astrophysique et Planétologie, Université de Toulouse, Toulouse, France
e-mail: arouillard@irap.omp.eu

P. Zucca
LESIA - Observatoire de Paris, CNRS, 92190 Meudon, France
e-mail: pietro.zucca@obspm.fr

D. Lario
Applied Physics Laboratory, Johns Hopkins University, Laurel, MD 20723, USA
e-mail: david.lario@jhuapl.edu

J. Pomoell
Department of Physics, University of Helsinki, 00014 Helsinki, Finland
e-mail: jens.pomoell@helsinki.fi

O.E. Malandraki
National Observatory of Athens, IAASARS, Athens, Greece
e-mail: omaland@astro.noa.gr

O.E. Malandraki, N.B. Crosby (eds.), *Solar Particle Radiation Storms Forecasting and Analysis, The HESPERIA HORIZON 2020 Project and Beyond*, Astrophysics and Space Science Library 444, DOI 10.1007/978-3-319-60051-2_9

We address two γ-ray events: 2012 January 23 and 2012 May 17 and approach the problem by, first, simulating the proton acceleration at the shock and, second, simulating their transport back to the Sun.

9.1 Introduction

The novel γ-ray observations by the Large Area Telescope (LAT) on the *Fermi* γ-ray Space Telescope spacecraft (Atwood et al. 2009), taken in a systematic way at unprecedented high energies, have presented a puzzle to the solar energetic particle (SEP) research community. More than two dozen >100 MeV γ-ray events were observed between 2008 and 2016 (see Chap. 8), many of which have properties that challenge the traditional idea that high-energy (>300 MeV) protons needed for the production of the γ-rays, via the pion-decay process deep in the chromosphere, are accelerated in solar flares (Ackermann et al. 2014). Specifically, the *Fermi*/LAT observations indicate that particles are precipitating to the solar atmosphere for up to a day after the impulsive phase of the flare, which is difficult to reconcile with a model of impulsive acceleration followed by particle trapping in the coronal magnetic field. On the other hand, coronal mass ejection (CME) driven shock waves can emit protons with energies above 300 MeV for several hours after the onset of the associated solar eruption as observed at 1 AU (e.g., the 2012 May 17 SEP event, see Chap. 8). Therefore, as an alternative view on the genesis of the long-duration γ-ray events, shock acceleration needs to be considered.

One of the challenges of the shock-acceleration hypothesis is that the SEP events observed in connection with the *Fermi*/LAT γ-ray events are not always very large, nor do they extend to very high energies when observed at 1 AU (see Chap. 8). Therefore, one of the key aspects to understand about these events is the spatial distribution of the accelerated particles at the CME-driven shock wave as well as the relation between interplanetary and interacting protons. Several factors contribute to this relation: (1) In-situ observations are local, i.e., performed in a particular interplanetary flux tube, whereas the observed high-energy γ-rays are produced over an extended emission region involving contributions from different field lines. (2) The energy spectrum of the particles accelerated at the shock is modified by transport effects when the particles propagate both downstream and upstream to reach the Sun and 1 AU. (3) Particles can modify their own transport conditions upstream of the shock due to amplification of Alfvén waves, so the fluxes observed at 1 AU can be partially decoupled from the fluxes at the shock. (4) Compressive and stochastic acceleration in the downstream region close to the CME can modify the spectrum of particles propagating toward the Sun.

We tackle the problem by conducting simulations of acceleration of protons in the shock and of their transport to 1 AU and back to the Sun for two long-duration γ-ray events: 2012 January 23 and 2012 May 17, of which the latter is associated with a ground level enhancement (GLE) of SEPs. In what follows, we outline in Sect. 9.2 the modelling techniques applied, then we present simulation results in Sect. 9.3 and their discussion as well as conclusions in Sect. 9.4.

9.2 Model Description

In this section, we outline the Shock-and-Particle (SaP) and Coronal Shock Acceleration (CSA) simulation models used to infer the proton spectrum at the shock and describe in detail the DownStream Propagation (DSP) model developed in the frame of High Energy Solar Particle Events foRecastIng and Analysis (HESPERIA) project and used to simulate proton transport from downstream of the shock to the Sun's surface.

9.2.1 *Shock and Particle Model*

The Shock-and-Particle (SaP) simulation model allows one to determine the injection rate of energetic particles from a propagating shock, using simulations of the shock propagation and particle transport, combined with the fitting of the simulation results to the observations. The particle injection rate Q at the shock is defined as $Q = \mathrm{d}f(z_\mathrm{s}, E, t)/\mathrm{d}t$, where f is the particle phase-space distribution function and $z_\mathrm{s}(t)$ is the shock position along the magnetic field line, which is, of course, a function of time. In SaP, the Parker spiral magnetic field is assumed. Instead of Q, one can consider the injection rate integrated over the cross-section A of the magnetic tube, $G = QA(z_\mathrm{s})$. See Chap. 4, Pomoell et al. (2015) and Aran et al. (2007) for further details.

9.2.2 *Coronal Shock Acceleration Model*

The Coronal Shock Acceleration (CSA) model is a Monte Carlo simulation model dealing with acceleration of ions in a coronal shock, taking into account ion-induced generation of Alfvén waves in the solar wind upstream of the shock. CSA simulates evolution of particles and Alfvén waves on a single radial magnetic field line. The shock is treated as a magnetohydrodynamic (MHD) discontinuity, the gas and magnetic compression ratios of which are computed through Rankine-Hugoniot relations, using the shock speed along the field line V_s, the shock-normal angle θ_Bn (the angle between the magnetic field vector $\mathbf{B}$ and the shock normal $\mathbf{n}$) and the ambient solar wind parameters (the plasma density n, the magnetic field B and the temperature T) at the shock position. All these parameters vary as the shock propagates outward from the Sun, which is implemented in CSA by using analytic functions of time or heliocentric distance to describe such variations. The analytic functions are determined by a number of free parameters that have to be provided as input in a simulation (see Afanasiev et al. 2017 for details). For instance, the variation of the magnetic field strength B with heliocentric distance r is implemented in CSA as

$$B(r) = B_0 \left(\frac{r_\oplus}{r}\right)^2 \left[1 + b_{\rm rf} \left(\frac{R_\odot}{r}\right)^6\right] , \tag{9.1}$$

where $R_\odot$ is the solar radius, $r_\oplus = 1$ AU, and B_0 and $b_{\rm rf}$ are free parameters. The parameter $b_{\rm rf}$ accounts for a super-radial expansion of the associated radial magnetic flux tube close to the solar surface. In HESPERIA, the free parameters were determined by fitting the analytic functions used to the data obtained by using techniques of semi-empirical modelling of the shock (see Rouillard et al. 2016).

The treatment of wave-particle interactions in CSA is based on quasi-linear theory, assuming only outward-propagating Alfvén waves, i.e., waves propagating away from the Sun. Particles experience elastic pitch-angle scattering in the wave frame, which is governed by the quasi-linear pitch-angle diffusion coefficient:

$$D_{\mu\mu} = \frac{\pi}{4} \Omega \frac{f_{\rm res} P(f_{\rm res})}{B^2} \left(1 - \mu^2\right) , \tag{9.2}$$

where $P(f)$ is the wave power spectrum, $f_{\rm res} = \Omega V/(2\pi v)$ is the resonant wave frequency, v and μ are the particle speed and pitch-angle cosine as measured in the wave-rest frame, V is the Alfvén wave propagation speed as measured in the solar-fixed frame and Ω is the ion cyclotron frequency. The expression given for the resonance frequency $f_{\rm res}$ represents a simplified (pitch-angle-independent) resonance condition of particle pitch-angle scattering (see, e.g. Afanasiev et al. 2015). If considered in other reference frames, in particular in the plasma frame, pitch-angle scattering leads to the energy exchange between particles and waves. Pitch-angle scattering of particles off outward-propagating waves, as viewed in the plasma frame, can give rise to growth of waves. After a series of approximations (Vainio 2003), the wave growth is given as

$$\Gamma(f_{\rm res}) = \frac{\pi}{2} \Omega \frac{p_{\rm res} S(r, p_{\rm res}, t)}{n V_{\rm A}} , \tag{9.3}$$

where $S(r, p, t) = 2\pi \int_{-1}^{+1} p^2 \mu v f(r, p, t) d\mu$ is the particle streaming, $p_{\rm res}$ is the resonant particle momentum corresponding to the resonance frequency $f_{\rm res}$, n is the proton density and $V_{\rm A}$ is the Alfvén speed.

Particles are followed in the guiding-center approximation in the shock's upstream. The seed particle population is modelled by a kappa distribution in speed with an exponential cutoff energy E_0. The spatial distribution of seed particles is parameterised as $n_{\rm seed} = \epsilon n$, where $n_{\rm seed}$ is the density of seed particles and ϵ is the injection parameter. Particle injection at the shock as well as particle-shock interactions are modelled by testing particles hitting the shock from the upstream for reflection/transmission from/through the shock front and for their transport back to the upstream, if transmitted. Alfvén waves are followed in the Wentzel-Kramers-Brillouin (WKB) approximation, with an additional diffusion term in frequency accounting for wave energy cascading. For a comprehensive description of CSA see Battarbee (2013).

9.2.3 DownStream Propagation Model

The DownStream Propagation (DSP) model is a Monte Carlo simulation model that has been devised to simulate proton transport from the shock's downstream back to the Sun. It is based on Parker's equation, which assumes quasi-isotropic particle distributions and, hence, diffusive transport:

$$\frac{\partial f}{\partial t} + \mathbf{u} \cdot \nabla f - \frac{1}{3} p (\nabla \cdot \mathbf{u}) \frac{\partial f}{\partial p} = \nabla \cdot (\kappa \cdot \nabla f) , \tag{9.4}$$

where $f = \langle \mathrm{d}^6 N / (\mathrm{d}^3 x \, \mathrm{d}^3 p) \rangle$ is the isotropic part of the particle distribution function, $\mathbf{u}$ is the velocity of the background fluid (solar wind plasma) and κ is the diffusion tensor. The second term on the left-hand side describes advection of particles with the solar wind, the third term describes adiabatic cooling due to the solar wind expansion and the right-hand side term describes spatial diffusion of particles. Hence, DSP describes the propagation of particles in the test-particle approximation.

Assuming that particles are confined within a magnetic flux tube, which gives $\kappa = \kappa \mathbf{bb}$ with $\mathbf{b}$ being a unit vector along the magnetic field, and that $\mathbf{u} \,||\, \mathbf{b}$, Parker's equation can be reduced to

$$\frac{\partial f}{\partial t} + u \frac{\partial f}{\partial z} - \frac{1}{3} p \frac{1}{A} \frac{\partial (Au)}{\partial z} \frac{\partial f}{\partial p} = \frac{1}{A} \frac{\partial}{\partial z} \left(A \kappa \frac{\partial f}{\partial z} \right) , \tag{9.5}$$

where z is the (curvilinear) coordinate measured from the solar surface along the field and $A(z)$ is the cross-sectional area of the flux tube. Changing to $F = 4\pi p^2 A f$, one can obtain

$$\frac{\partial F}{\partial t} + \frac{\partial}{\partial z} \left[\left(u + \frac{1}{A} \frac{\partial (A\kappa)}{\partial z} \right) F \right] - \frac{\partial}{\partial p} \left(\frac{1}{3} p \frac{1}{A} \frac{\partial (Au)}{\partial z} F \right) = \frac{1}{2} \frac{\partial^2}{\partial z^2} (2\kappa F) . \tag{9.6}$$

Equation (9.6) is equivalent to the following set of stochastic differential equations (SDEs):

$$\mathrm{d}z = \left(u + \frac{\kappa}{L} + \frac{\partial \kappa}{\partial z} \right) \mathrm{d}t + \sqrt{2\kappa} \, \mathrm{d}W_t , \tag{9.7}$$

$$\mathrm{d}p = -\frac{1}{3} p \left(\frac{u}{L} + \frac{\partial u}{\partial z} \right) \mathrm{d}t , \tag{9.8}$$

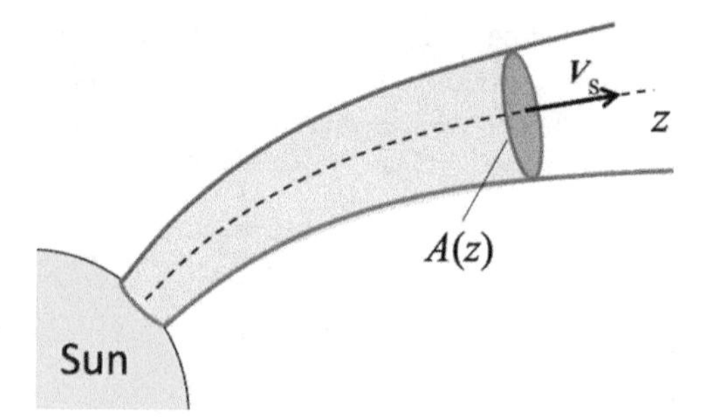

Fig. 9.1 Spatial simulation domain in the DSP model

where $L(z) = A/(\mathrm{d}A/\mathrm{d}z)$ is the focusing length of the magnetic flux tube, $\mathrm{d}W_t$ is a stochastic differential normally distributed with zero mean and variance $\mathrm{d}t$. Note that the term $(\kappa/L)\mathrm{d}t$ in Eq. (9.7) represents the net effect of focusing to a quasi-isotropic particle population (Kocharov 1996). These SDEs are solved for each Monte Carlo particle in the simulation, using the standard explicit Euler-Maruyama method.

The simulation is performed in an expanding 1-D spatial simulation box in the shock's downstream, i.e., the box is confined by the solar surface from one side and by the moving shock front from the other side, and, thus, expands with time along the magnetic field line (Fig. 9.1). For the plasma speed u and the spatial diffusion coefficient κ in the simulation box, the following linear models are adopted:

$$u(z,t) = \frac{u_0 z}{V_\mathrm{s} t}, \tag{9.9}$$

$$\kappa(z,p,t) = \frac{1}{3}\lambda_0 \frac{p}{m_\mathrm{p}} \frac{V_\mathrm{s} t - z}{V_\mathrm{s} t}, \tag{9.10}$$

where u_0 is the plasma speed immediately behind the shock front and V_s is the shock speed along the field, both measured in the solar-fixed frame, m_p is the proton mass and λ_0 is the particle mean free path at the solar surface. One can see from Eqs. (9.9) and (9.10) that the plasma speed decreases linearly from $u = u_0$ at the shock ($z = V_\mathrm{s} t$) to $u = 0$ at the Sun ($z = 0$), and the mean free path recovers from $\lambda = 0$ at the shock to $\lambda = \lambda_0$ at the Sun. Hence the transport of particles from the upstream through the shock is purely advective. On the other hand, the linear dependence of the spatial diffusion coefficient κ on $V_\mathrm{s} t - z$ secures that no particles escape from the downstream beyond the shock.

The focusing length L is specified in a form similar to the one that can be derived from Eq. (9.1):

$$L(z) = \frac{1}{2}(R_\odot + z)\left[\left(1 + \frac{z}{R_\odot}\right)^6 + b_\mathrm{f}\right]\left[\left(1 + \frac{z}{R_\odot}\right)^6 + 4b_\mathrm{f}\right]^{-1}, \tag{9.11}$$

where b_f is a free parameter.

The initial size of the simulation box is specified by the initial position of the shock z_{s0}. This gives the initial value for time t as $t_0 = z_{s0}/V_s$. The shock speed V_s is assumed constant and $u_0 = V_s(r_c - 1)/r_c$ with r_c being the gas compression ratio taken constant as well.

To be consistent with the chosen form of κ, the amount of particles being injected into the simulation from the shock is determined by the net particle flux to the far downstream. This flux is given by

$$\mathscr{F} = \int d^3p\mu v f(p,\mu) = \int d^3p' u_2 f'(p') , \tag{9.12}$$

where μ is the pitch-angle cosine, v is the particle speed, u_2 is the downstream solar wind speed; the unprimed symbols designate quantities as measured in the shock frame and the primed symbols in the downstream plasma frame. The second expression in Eq. (9.12) is derived under the assumption that the particle pitch-angle distribution as measured in the downstream plasma frame is isotropic. Then, it can be derived to the first order in u_2/c (c is the speed of light in vacuum), taking into account that $\lambda \to 0$ at the shock, that

$$\frac{d\mathscr{F}}{dp} = 4\pi u_2 p^2 f_0(p) , \tag{9.13}$$

where $f_0(p)$ is the isotropic part of the particle distribution function at the shock (on the upstream side). This equation is used to relate the amount of particles injected to the downstream with the particle intensity at the shock, $j_s = p^2 f_0(p)$, which is the output of CSA simulations. Specifically, the number of particles per unit momentum injected downstream from a source (at the shock) of size $A_s(z)$ in time dt is determined by

$$\frac{dN}{dp} = 4\pi \frac{V_s}{r_c} A_s(z) j_s(t,E) dt . \tag{9.14}$$

In case the particle injection rate $G(t,E)$ is given (output of SaP), the particle intensity at the shock is computed as

$$j_s(t,E) = \frac{p^2}{v} \frac{C \Delta r_g G(t - t_0, E)}{r^2 \cos\psi(r)} , \tag{9.15}$$

where v is the particle speed, $C = 3.524$ is a constant determined by the numerical scheme implemented in SaP, Δr_g is the spatial grid size in SaP simulations, r is the heliocentric distance and it is taken into account that in SaP the injection rate G is determined under the assumption of a Parker-spiral upstream magnetic field, i.e., $\cos\psi(r) = [1 + (\Omega_\odot r/u_{sw})^2]^{-1/2}$ with constant solar wind speed u_{sw} ($\Omega_\odot$ is the angular speed of the solar rotation).

The proton injection into the downstream is implemented in the following way. We introduce an injection time step Δt and deposit a certain number of Monte Carlo particles into the simulation box at each Δt. A Monte Carlo particle is a representative of a group of physical particles and is characterised by its weight w. To obtain weights of injected Monte Carlo particles at a given time t_0, we compute values of the spectrum $\mathrm{d}N/\mathrm{d}p$ at the momentum grid using Eq. (9.14) and interpolate those spectral values by power laws between the grid nodes. Then, the weight of a Monte Carlo particle with momentum p between the grid points p_j and p_{j+1} is given by

$$w = \frac{1}{N_{\mathrm{MC}}} \int_{p_j}^{p_{j+1}} \frac{\mathrm{d}N}{\mathrm{d}p} \mathrm{d}p, \tag{9.16}$$

where $\mathrm{d}N/\mathrm{d}p \propto p^{-q_j}$ and N_{MC} is the number of Monte Carlo particles assigned to the interval $[p_j, p_{j+1}]$. Note that the particle momentum p is randomly chosen from a p^{-q_j} distribution in this interval and $q_j = -\ln(S_{j+1}/S_j)/\ln(p_{j+1}/p_j)$ with $S_{j(j+1)} = \mathrm{d}N/\mathrm{d}p|_{j(j+1)}$.

The Monte Carlo particles injected at a given time t_0 are placed into a small spatial 1-D volume $\Delta z = u_2 \Delta t$ behind the front, to mimic their advection by the bulk plasma during Δt. Then, each of these particles is propagated by solving the SDEs until it hits the Sun, i.e., precipitates, or up to $t = t_{\max}$. The particle transport time step δt must fulfil the condition $\delta t \ll L^2/\kappa$. Furthermore, the linear dependence of κ on distance from the shock dictates that Monte Carlo particles that might appear in the upstream side during a simulation have to be returned to the downstream side. This requires a reflective boundary condition for particles to be applied at the shock.

During the simulation, Monte Carlo particles that hit the Sun are collected and, based on their momenta, precipitation times and weights, the flux and time-integrated energy spectrum of absorbed protons are calculated. The latter is compared with the proton spectra derived from γ-ray observations.

9.3 Results

The simulations based on the models presented above were conducted for two long-duration >100 MeV γ-ray events occurred on 2012 January 23 and 2012 May 17. These γ-ray events are associated with substantial SEP events as observed at 1 AU (the 2012 May 17 event is also a GLE). The event characteristics are described in Chap. 8 (see also Rouillard et al. 2016). In what follows we first present modelling results for the 2012 May 17 event due to its association with a GLE and then results for the 2012 January 23 event.

9.3.1 2012 May 17 Event

9.3.1.1 Modelling of the SEP Event

We begin with results of the SaP modelling of the accelerated proton population at the shock, which utilises the observations of the shock and SEP event at 1 AU. Figure 9.2 presents results of simulations of the ambient solar wind and shock propagation to the Earth. The solar wind was simulated starting from the onset time of the SEP event. Table 9.1 provides values of the input solar wind parameters at 1.03 $R_{\odot}$. The input parameters of the shock-driving disturbance are its density $\rho_{\rm cme} = 0.3 \times 10^{-13}$ kg m^{-3}, speed $v_{\rm cme} = 1000$ km s^{-1}, and the shock-front shape parameter $\Delta\phi = 0.5 a_{\rm cme}$ determined by the angular extent $a_{\rm cme}$ of the disturbance (see Pomoell et al. 2015 for details). One can see that the simulations reproduce quite well the average characteristics of the solar wind at 1 AU prior to the shock arrival (note, however, that the temperature is somewhat underestimated) as well as the shock arrival time. At the same time, the observed jumps in the plasma characteristics are reproduced well with some underestimation of the magnetic field increase.

In order to derive the proton injection rate from the shock into the magnetic flux tube connecting the observer with the shock front, we fitted the proton

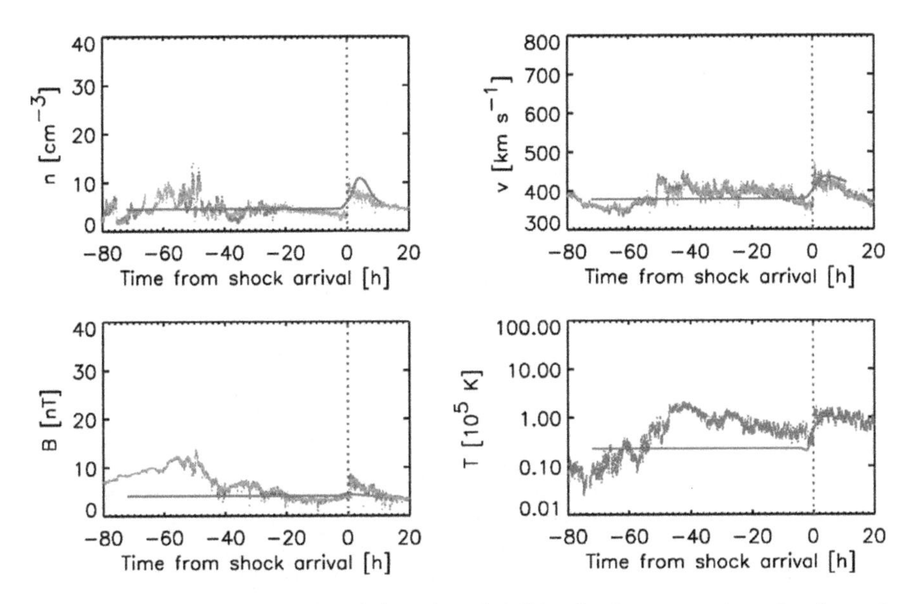

Fig. 9.2 Simulated characteristics of the solar wind (*blue lines*) superposed on the observed characteristics (shown in *orange and red*) before and during the shock arrival at 1 AU for the 2012 May 17 SEP event. Shown are: the plasma density, the solar wind radial velocity, the magnetic field magnitude and the solar wind proton temperature. The temperature was measured by the Wind spacecraft and the other parameters by the ACE spacecraft. Time counts from the time of the interplanetary shock passage by the ACE spacecraft (marked by a *dotted vertical line*)

Table 9.1 Input solar wind parameters used in the SaP simulations of the 2012 January 23 and 2012 May 17 SEP events

Event	ρ_0 (kg/m^3)[a]	T_0 (K)	S_0 (W/m^3)	$\mathscr{L}$ ($R_\odot$)	γ	B_{r0} (T)
Jan 2012	1.169×10^{-13}	1.22×10^6	0.335×10^{-7}	0.735	1.55	-2.15×10^{-4}
May 2012	1.169×10^{-13}	1.18×10^6	0.35×10^{-7}	0.70	1.55	1.26×10^{-4}

[a]The parameters provided are the plasma density ρ_0, temperature T_0 and the radial component of the magnetic field B_{r0} at the heliocentric distance $r_0 = 1.03\ R_\odot$; S_0 and $\mathscr{L}$ are the coronal heating function parameters (Pomoell et al. 2015); and γ is the adiabatic index.

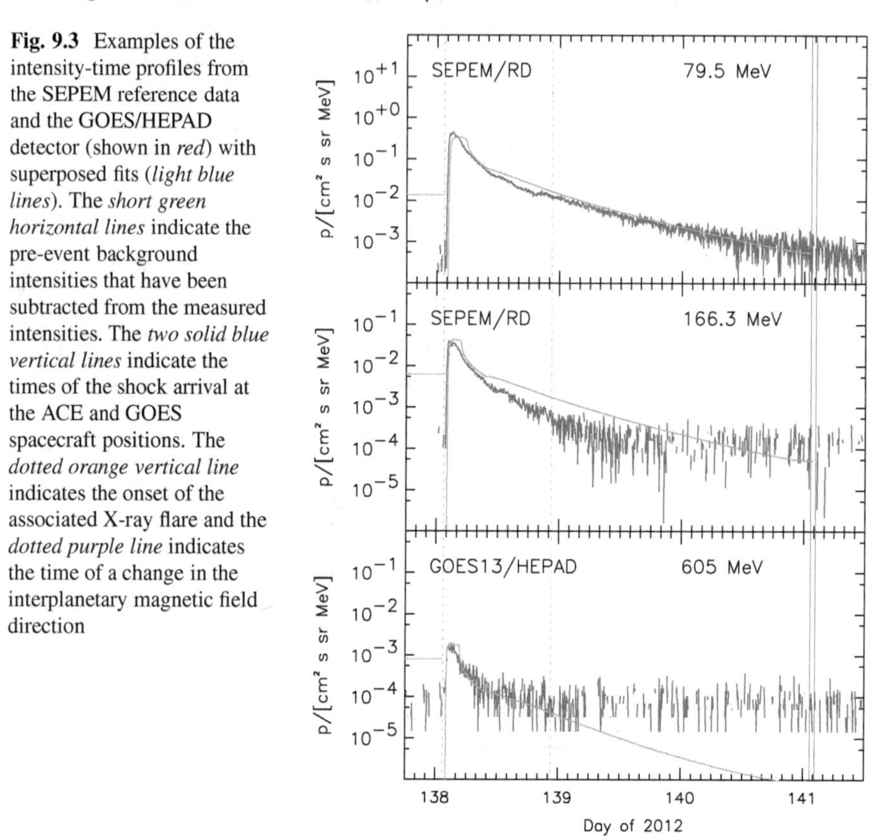

Fig. 9.3 Examples of the intensity-time profiles from the SEPEM reference data and the GOES/HEPAD detector (shown in *red*) with superposed fits (*light blue lines*). The *short green horizontal lines* indicate the pre-event background intensities that have been subtracted from the measured intensities. The *two solid blue vertical lines* indicate the times of the shock arrival at the ACE and GOES spacecraft positions. The *dotted orange vertical line* indicates the onset of the associated X-ray flare and the *dotted purple line* indicates the time of a change in the interplanetary magnetic field direction

intensity-time profiles provided by the ACE/EPAM instrument (in the energy range 0.59–4.8 MeV), SEPEM reference data (Jiggens et al. 2012) (6–166.3 MeV) and GOES/HEPAD detector (330–700 MeV). The fitting was performed using SaP particle transport simulations, following the method described in Pomoell et al. (2015). Figure 9.3 shows examples of the fitted intensity-time profiles for several high-energy channels. Note that when performing the fitting, we focused on the first several hours of the event as the corresponding γ-ray event lasted for about 2 h.

Figure 9.4 shows the proton injection rates $G(t)$ resulting from the fitting of the intensity-time profiles at selected energies. The obtained injection rates were used

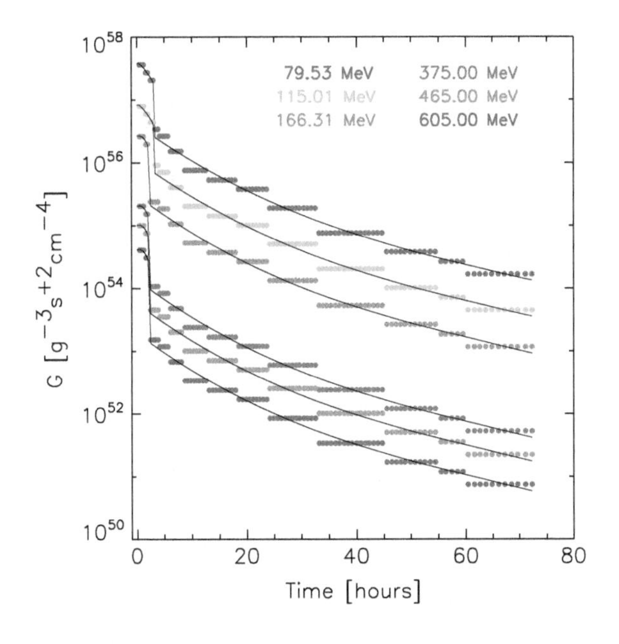

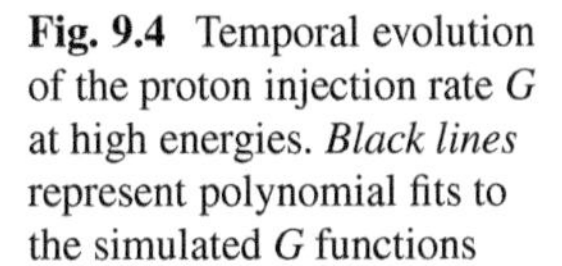
Fig. 9.4 Temporal evolution of the proton injection rate G at high energies. *Black lines* represent polynomial fits to the simulated G functions

as input for the DSP simulations. The simulations reveal that the injection rate at high energies drops by more than one order of magnitude several hours after the X-ray flare onset, which agrees with rather short duration of the associated γ-ray event ($\sim$2 h).

9.3.1.2 Simulations of Proton Acceleration at the Shock

To carry out the CSA simulations of proton acceleration at the shock in this event, we utilised the ambient plasma and shock parameters derived from the semi-empirical modelling of the shock (Rouillard et al. 2016). Those parameters are the plasma density n, the magnetic field strength B, the shock speed V_s and the shock-normal angle θ_{Bn}, which were determined along individual magnetic field lines. The parameters were fitted by the analytic functions of time/distance implemented in CSA (Afanasiev et al. 2017). Figure 9.5 shows an example of the data obtained for a single field line, superposed by the corresponding fits. In total, the data for over 100 field lines were available and fitted. Among the field lines possessing good-quality fits, nine were chosen for simulations with CSA.

The detailed analysis of the simulation results presented in Afanasiev et al. (2017) reveals that the parameter mainly controlling the acceleration efficiency is the Alfvénic Mach number of the shock. Figure 9.6 shows examples of the evolution of the Alfvénic Mach number of the propagating shock for three different magnetic field lines and simulated proton energy spectra at the shock, corresponding to these field lines, obtained at $t = 1000$ s. The simulated spectra (for one of the simulated field lines) were then used as the other input for the DSP simulations.

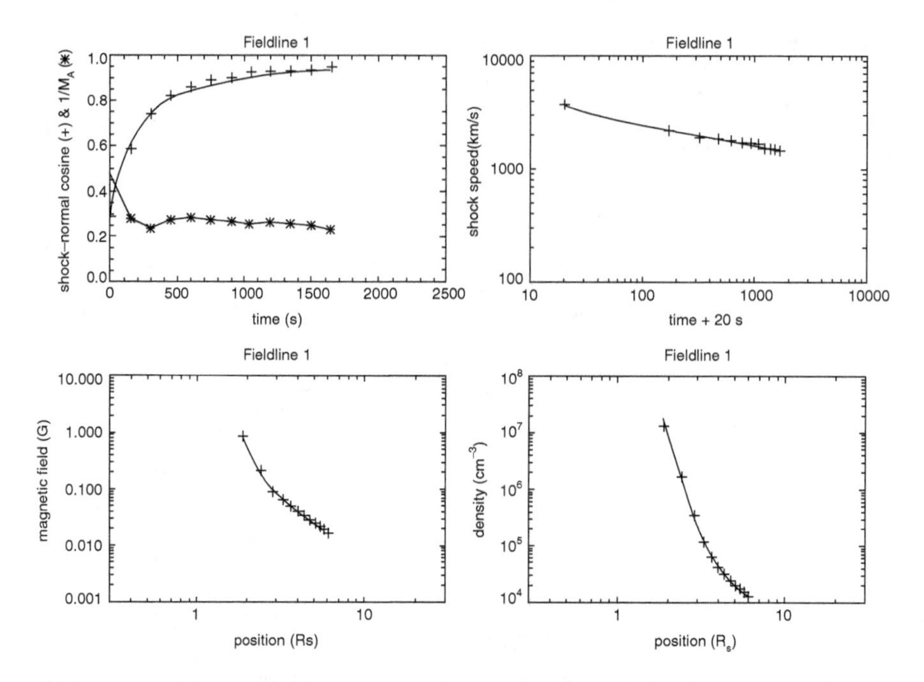

Fig. 9.5 Example data set obtained with the semi-empirical shock modelling approach in the 2012 May 17 event, plotted together with the corresponding fits. The *upper left panel* shows the shock-normal cosine cos θ_{Bn} ("plus" symbols) and the inverse Alfvénic Mach number M_{A}^{-1} (asterisks) vs. time; the *upper right panel* shows the shock speed V_{s} along the magnetic field line vs. time. Note that time is counted from the moment when $M_{\mathrm{A}} > 1.5$. The *bottom left panel* shows the magnetic field magnitude B vs. radial shock position and the *bottom right panel* shows plasma density n vs. radial shock position

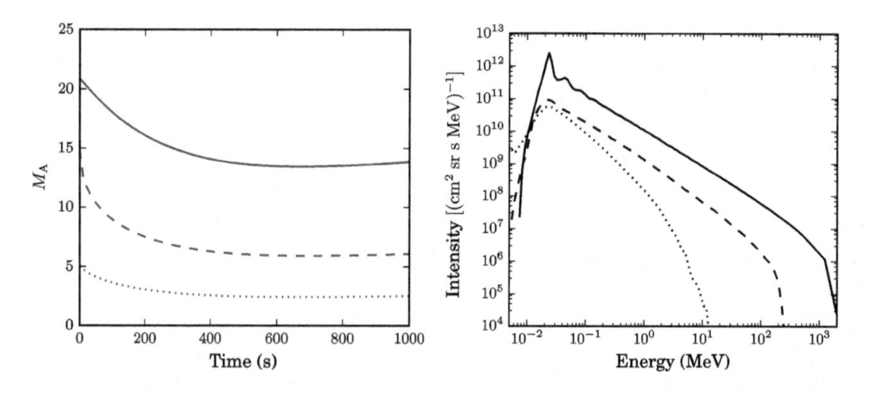

Fig. 9.6 Alfvénic Mach number of the shock versus simulation time for three different magnetic field lines (*left panel*) and corresponding proton energy spectra at the shock at $t = 1000$ s, resulting from CSA simulations (*right panel*). Note the correspondence between the Alfvénic Mach number and the spectral cutoff energy

9.3.1.3 Modelling of the Proton Transport Back to the Sun

The proton transport back to the Sun was simulated with the DSP model, assuming a radial magnetic field in the shock's downstream. The particle source size at the shock was modelled as $A_s(z) = (R_\odot + z)^2\Omega_0$. The parameter Ω_0, which can be interpreted as the global angular size of the shock, was taken to be 1 steradian. In fact, the realistic effective source size should be smaller because of the substantial difference in the particle acceleration efficiency along different field lines, as revealed by the CSA simulations for this event. We took this into account by considering an additional parameter, so-called filling factor, $a_{\rm fill}$ that is the relative fraction of field lines at which high-energy, γ-ray-productive, protons can be produced. Based on the ascertained dependence between the particle acceleration efficiency and the Alfvénic Mach number magnitude, and available data on >100 individual field lines, we estimated that $a_{\rm fill} = 0.1$. To compare the proton spectra resulting from the DSP simulations with the proton spectrum obtained from the *Fermi*/LAT observation, we multiply the simulated spectra by $a_{\rm fill}$. The parameters of the observationally-derived spectrum (the total number of >500 MeV protons and the power-law spectral index of >300 MeV protons) were kindly provided to us by G. Share (see Chap. 8 for references). The other DSP model parameters were taken to be $z_{s0} = 1.6\,R_\odot(1.15\,R_\odot)$, in the case of CSA(SaP) input, $V_s = 1510$ km s^{-1}, $r_c = 3.6$ and $u_{sw} = 387$ km s^{-1}.

The DSP simulations were conducted for a set of values for the downstream transport parameters λ_0 and b_f. The resulting (integrated over the duration of the γ-ray emission; i.e., 2 h) energy spectra of protons hitting the Sun are presented in Fig. 9.7 along with the shock-injected spectrum and the observationally-derived

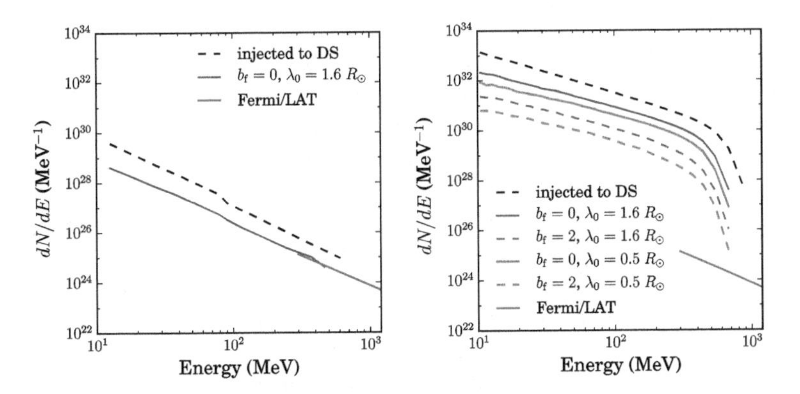

Fig. 9.7 Time-integrated energy spectra of protons precipitated at the Sun, resulting from DSP simulations using the results of SaP (*left panel*) and CSA simulations (*right panel*) for the 2012 May 17 event. The spectra are shown by *blue and green lines* with the corresponding DSP model parameters indicated. The integration time is 2 h, which is approximately the duration of the >100 MeV γ-ray event. Also shown are the time-integrated spectrum of protons injected by the shock to the downstream (*black dashed line*) and the spectrum of interacting protons derived from the *Fermi*/LAT observation (*red line*)

spectrum of interacting >300 MeV protons. It can be calculated that the number of >500 MeV protons injected from the shock in the CSA+DSP simulation exceeds by more than 10^4 the corresponding number of protons derived from the observation. On the other hand, the number of absorbed high-energy protons is sensitive to the transport parameters. In particular, it can be easily reduced by increasing the parameter b_f controlling the focusing length, which enhances proton mirroring from the flux tube base and adiabatic cooling, in accord with Eqs. (9.7) and (9.8). Note also that the DSP model completely neglects the possibility for particles in the downstream side of the shock to escape to the upstream side. This process, if taken into account, should decrease the number of absorbed particles as well.

In contrast, the spectrum of shock-injected protons obtained from the SaP+DSP simulation and the absorbed spectrum corresponding to $b_f = 0$ (radial flux tube) are in good correspondence with the observed spectrum. However, it should be noted that this correspondence holds only for rather idealistic conditions of the DSP model (no particle escape to the upstream) and one can expect a lack of high-energy protons in the simulations, if a more realistic downstream transport model is considered. The possible reasons of this result are discussed in Sect. 9.4.

9.3.2 2012 January 23 Event

9.3.2.1 Modelling of the SEP Event

The results of simulations of the ambient solar wind and shock propagation to 1 AU are shown in Fig. 9.8. The input parameters describing the initial shock-driven disturbance in this case are $\rho_{\rm cme} = 0.3 \times 10^{-13}$ kg m^{-3}, $v_{\rm cme} = 1650$ km s^{-1}, and $\Delta\phi = 0.25 a_{\rm cme}$ (see also Table 9.1 for the initial solar wind parameters). Like for the 2012 May 17 event, the modelling reproduces well the average characteristics of the solar wind at 1 AU prior to the shock arrival and the shock arrival time. The observed jump in the magnetic field magnitude is reproduced well too, but the jumps in the density, the solar wind speed and the temperature are somewhat overestimated. Note that due to the data gap in the ACE data, the value of the average temperature (2.5×10^4 K) in the upstream region has been taken from the plot provided by the IP shocks data base of the University of Helsinki,[1] based on the WIND data. A similar value is estimated by the CfA interplanetary shock list.[2]

Like for the previous event, we fitted the observed intensity-time profiles provided by the ACE/EPAM instrument and the SEPEM reference data (Jiggens et al. 2012). The proton enhancements observed in this event by GOES/HEPAD in the high-energy channels ranging from 330 to 700 MeV are weak. From Fig. 9.9, it can be seen that already at 166.3 MeV the background-subtracted enhancement is lower than the background level itself. For this reason, instead of fitting the observed

[1]http://ipshocks.fi/database.

[2]https://www.cfa.harvard.edu/shocks.

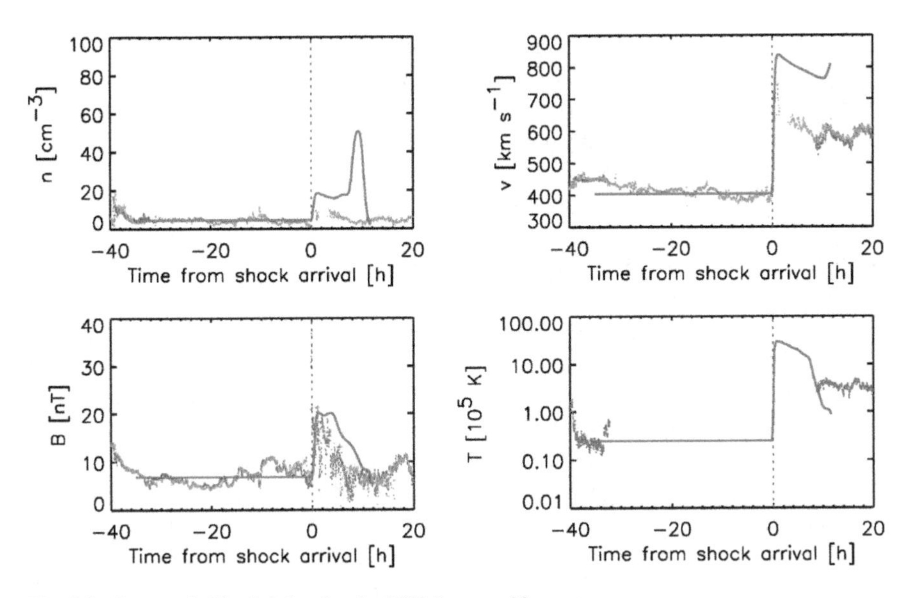

Fig. 9.8 Same as in Fig. 9.2, but for the 2012 January 23 event

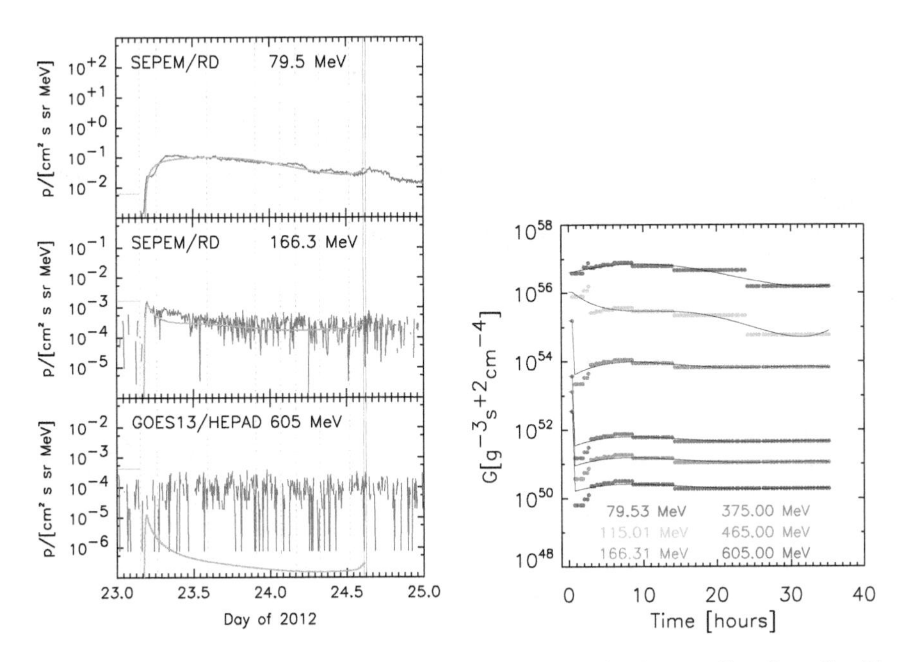

Fig. 9.9 *Left panel*: Examples of the observed proton intensity-time profiles (*in red*) with superposed fits/synthetic profiles (*light blue lines*) for the 2012 January 23 SEP event. The *short green horizontal lines* indicate the pre-event background intensities that have been subtracted from the measured intensities. The meaning of the *other lines* is the same as in Fig. 9.3. *Right panel*: Temporal evolution of the proton injection rate G at high energies. *Black lines* are polynomial fits to the simulated G functions

intensities, we computed synthetic intensity-time profiles for the GOES/HEPAD energy channels by extrapolating the particle injection rate $G(t, E)$ from the highest energies in the SEPEM reference dataset. Such a profile is shown in Fig. 9.9 at 605 MeV. Figure 9.9, right panel shows the proton injection rate resulting from the SaP simulations of this SEP event.

9.3.2.2 Simulation of Proton Acceleration at the Shock

To conduct CSA simulations of proton acceleration at the shock in the same fashion as it was done for the 2012 May 17 event, we used white-light (WL) images of the corona from SOHO and STEREO A and B and the Potential Field Source Surface (PFSS) modelling of the magnetic field. The WL images were fitted using a spherical representation for the CME. This allowed us to obtain the magnetic field, the shock speed and the shock-normal angle along different field lines (Fig. 9.10). As concerns the plasma density, we used the following representation:

$$n(r) = n_2 \left(\frac{r_\oplus}{r}\right)^2 + n_8 \left(\frac{R_\odot}{r}\right)^8, \tag{9.17}$$

with $n_2 = 10\ \text{cm}^{-3}$ and $n_8 = 8 \times 10^8\ \text{cm}^{-3}$. Since the plasma density was not constrained using the real observations, we performed a CSA simulation only for one magnetic field line. The simulation showed a typical buildup of the proton energy with time and formation of a power-law energy spectrum with a roll-over, similar to the spectra presented in Fig. 9.6. The maximum proton energy at the shock attained the value ∼1 GeV at $t = 1130$ s after the start of the simulation, but then was decreasing and reached ∼500 MeV at $t = 2000$ s.

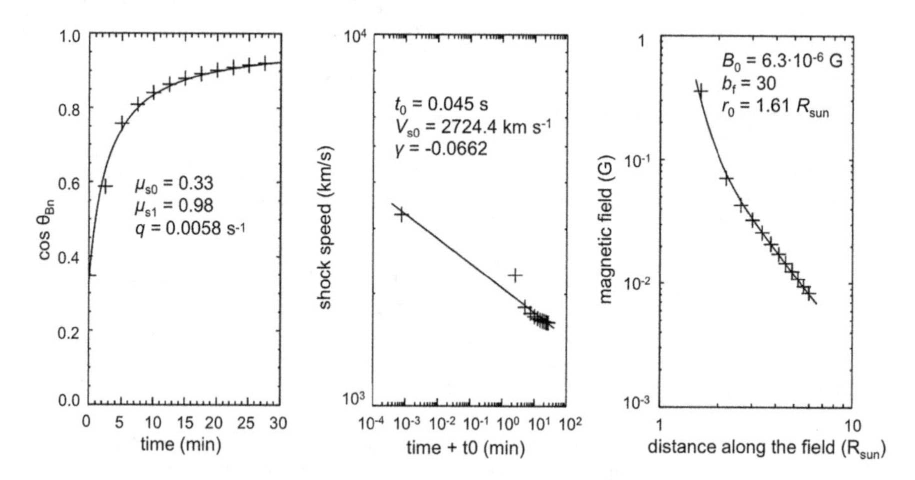

Fig. 9.10 The shock-normal cosine μ_s, the shock speed along the field line V_s and the magnetic field B with the fits superimposed for a selected magnetic field line

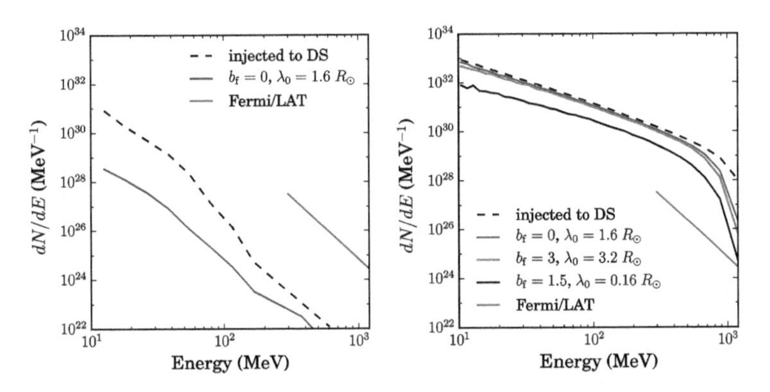

Fig. 9.11 Time-integrated energy spectra of protons precipitated at the Sun, resulting from DSP simulations using the results of SaP (*left panel*) and CSA simulations (*right panel*) for the 2012 January 23 event. The spectra are shown by *blue and green lines* with the corresponding DSP model parameters indicated. The integration time is 6 h. Also shown are the time-integrated spectrum of protons injected by the shock to the downstream (*black dashed line*) and the proton spectra obtained from the *Fermi*/LAT observation (*red line*)

9.3.2.3 Modelling of the Proton Transport Back to the Sun

Similar to the DSP modelling for the 2012 May 17 event, we assumed a radial magnetic field in the shock's downstream and $A_s(z) = (R_\odot + z)^2\Omega_0$ was taken for the particle source at the shock. The model parameters were taken to be $z_{s0} = 0.6\,R_\odot(3.1\,R_\odot)$ in the case of CSA(SaP) input, $V_s = 1450$ km s^{-1}, $r_c = 3.7$, $u_{sw} = 414$ km s^{-1}, $\Omega_0 = 1$ sr, and $a_{fill} = 0.1$.

Figure 9.11 shows the simulated spectra of protons absorbed at the Sun together with the shock-injected spectrum and the proton spectrum derived from the *Fermi*/LAT observation of $>$100 MeV γ-photons. Similar to the simulation for the 2012 May 17 event, in the CSA+DSP simulation, there is a substantial excess of shock-injected protons as compared to the number of interacted protons derived from the observation. In contrast, there is a lack of shock-injected protons against the observed number in the SaP+DSP simulation.

9.4 Discussion and Conclusions

We have modelled particle acceleration at coronal shocks driven by CMEs and proton transport from the shock to both the Sun and the far upstream region (towards the 1-AU observer). The purpose of our study is to find out, whether shock-accelerated protons streaming back from the shock could be responsible for the long-duration γ-ray events observed by *Fermi*/LAT. We simulated the shock propagation from the Sun to 1 AU using a two-dimensional MHD model. We also employed the empirical models of the CME-driven pressure front propagation, which allowed us to assess the early evolution of the shock in a system that does not possess the symmetries assumed in the MHD model.

Our results show that the efficiency of particle acceleration crucially depends on the modelled properties of the shock in the corona. Conditions on different field lines vary very much and while the shock on some field lines is able to produce a relativistic particle event, it fails to do so on others. The most important factor governing the acceleration efficiency in our study was the Alfvénic Mach number of the shock: the higher the Alfvénic Mach number, the more likely the shock to accelerate protons to relativistic energies (Afanasiev et al. 2017).

In our simulations, we focused on two events that differ in one important aspect: one of them (2012 May 17) is a GLE and the other one is not. Both of the events are associated with long-duration γ-ray events, which might seem contradictory, but actually is not. In the light of our simulations, there are two possible explanations for this. Firstly, as the particle acceleration efficiency at the shock varies a lot from one flux tube to another, the 1-AU proton event does not necessarily correspond to the best acceleration conditions on the shock surface. Thus, we may well observe a long-duration (pion-decay) γ-ray event due to shock-accelerated protons without a clear increase observed at 1 AU at energies required for pion production at the Sun. The Earth-based observer sees but a small fraction of the complete picture. The other, more subtle explanation deals with the strength of the turbulence in the foreshock region of the coronal shock. The CSA simulations show that the foreshock is extremely turbulent near the Sun, traps a large fraction of particles (almost all) in its vicinity and allows only a minor fraction to escape. While the SaP model contains the possibility to use enhanced foreshock turbulence, it is tuned to reproduce the observations when the shock is detected in situ. Therefore, the source function deduced from the 1-AU observation represents more the fraction of particles that can escape upstream than the fraction that can be transported downstream from the shock. The large discrepancy between the CSA and SaP modelled spectra of precipitated particles is, thus, partly explained by this effect, as well. Furthermore, these explanations shed light on the tendency to get lower numbers of precipitated >300 MeV protons in the SaP+DSP simulations (especially for the 2012 January 23 event), as compared to the observations.

The transport model we employ for the downstream region has several important simplifications in it. Firstly, it employs a shock completely opaque to protons and, thus, allows downstream-advected particles to reside in the region between the shock and the Sun for as long as they get precipitated. The only loss process we employ is adiabatic cooling of the distribution, when the region between the shock and the Sun expands. On the other hand, we do not include any downstream re-acceleration processes, which could also be important and would act in the opposite direction, helping particles to overcome adiabatic energy losses. Such processes include downstream stochastic acceleration (see, e.g., Afanasiev et al. 2014) and compressive acceleration in the highly-compressed regions close to the CME core (Kozarev et al. 2013). Therefore, we do not regard our model to be overly optimistic about the prospect of letting shock-accelerated protons precipitate over large time scales.

Regarding the CSA model, one important reservation has to be made: the model makes a simplification to the quasi-linear resonance condition between particles and scattering Alfvén waves, $k_{\text{res}} = 2\pi f_{\text{res}}/V = B/(R\mu)$, neglecting the dependence of resonant wave number k_{res} on particle pitch-angle cosine μ (R is the particle rigidity). This simplification, while without proper physical justification, allows one to build the code using the assumption of isotropic scattering, which speeds up the running times easily by a factor of ten over times obtained when treating the resonance condition in full. We have, however, evaluated the effect of this simplification using a local model (Afanasiev et al. 2015), and shown that in parallel coronal shocks the difference between the two models yields about a factor of 2 in the roll-over rigidity obtained from the model. On the other hand, the spectrum in CSA also cuts off much more rapidly than in the model employing the complete resonance condition, so we do not regard this to be a very serious problem in the performed modelling. A more complete global model to be developed in future, however, should take the full resonance condition into account, also since it affects the foreshock spatial structure as well (Afanasiev et al. 2015). We will undertake the development of such a model in future projects.

Another transport process missing from the CSA model is diffusion perpendicular to the mean magnetic field. This process can be implemented in a Monte Carlo simulation, but its inclusion will require to incorporate at least one more spatial dimension in the model. Therefore, also the requirement for particle statistics in CSA will be tremendously increased to avoid statistical noise in the result, as the number of spatial cells in the model will have to be increased by a factor 30–300, depending on the coarseness of the grid in the perpendicular direction. This is still beyond the reach of the present computers with reasonable running times of the code. Fortunately, as perpendicular diffusion cannot occur due to slab-mode waves, the enhanced Alfvénic turbulence in the upstream region is not strengthening the perpendicular diffusion of the particles. However, one would expect the downstream plasma to have much more isotropic turbulence which, then, would lead to the migration of particles from one flux tube to the other while they are on the downstream side of the shock. For an opaque shock, like we have assumed, this is not affecting the acceleration of ions at the shock too much since their fate (scattered back to the shock or transported to the far downstream region with no return) would be decided (almost) instantly, giving the particles very little time to diffuse perpendicular to the mean field. Thus, the first step to take the perpendicular transport into account would have to be performed on the downstream side of the shock.

We note that the total number of >500 MeV protons as simulated by the CSA code is several orders of magnitude larger than the observational value in both of the simulated events. At first, this might seem to be problematic. However, in addition to the caveats about the resonance condition and the downstream transport modelling discussed above, we point out that the CSA model is set up quite favourably for efficient particle acceleration: we use a seed particle population with a relatively hard suprathermal distribution ($\kappa = 2$) in the model, which guarantees an efficient injection to the acceleration process at all obliquity angles of the shock. Using a thermal population, only, would limit the injection efficiency of the shock especially at obliquity angles greater than $\sim 20°$ quite substantially: according to Battarbee et al. (2013), the injection efficiency in an oblique coronal shock would go down by an order of magnitude if we would use a steeper seed population with $\kappa = 15$. Given all the possible ways to limit the number of precipitating protons in our model, we would regard the result of getting more than enough high-energy protons precipitating at the Sun to be a supporting rather than a countering result for the shock acceleration scenario.

The final shortcoming in the shock models we have employed is their assumption of the open topology of the upstream magnetic fields. Especially if the γ-ray event occurs during a period of closely spaced CMEs, the second one may drive a shock through a set of large closed loop-like or flux-rope structures, which would alter both the shock acceleration properties and, more importantly, the ability of the particles to escape upstream from the shock. In fact, developing codes capable of modelling particle acceleration and transport in more complicated upstream fields than the radial/Archimedean-spiral fields could be listed as one of the most urgent things needing improvement on the way towards physical space-weather modelling capabilities.

One of the most difficult things to estimate is the size of the source of near-relativistic protons in the event. While the 3-D modelling of the shock front can be performed in a relatively accurate and detailed manner, high-resolution density and magnetic-field structure of the corona are crucial for the correct determination of the shock properties and, thus, the total number of interacting protons in the event. Therefore, even a fully global 3-D model of coronal shock evolution and particle acceleration might not capture every detail affecting the total number of relativistic protons in the CME system. In this work, we resorted to estimating numbers based on the filling factor of field lines being capable of facilitating proton acceleration to relativistic energies based on an evaluation of shock properties on a large set of field lines. We believe that such statistical method to estimate the total number of interacting protons is the most efficient way to address the problem.

In conclusion, while a number of simplifications have been introduced in the modelling performed, we have still demonstrated that coronal shock acceleration and subsequent diffusive downstream particle transport is a viable option to explain pion-decay γ-ray events from the Sun observed by *Fermi*/LAT. More elaborated simulation models are needed (especially for the particle transport back to the Sun) but our results serve as a motivation by indicating that the end result of this vast modelling effort can be positive.

References

Ackermann, M., Ajello, M., Albert, A., et al.: Astrophys. J. **787**, 15 (2014). doi:10.1088/0004-637X/787/1/15

Afanasiev, A., Vainio, R., Kocharov, L.: Astrophys. J. **790**, 36 (2014). doi:10.1088/0004-637X/790/1/36

Afanasiev, A., Battarbee, M., Vainio, R.: Astron. Astrophys. **584**, A81 (2015). doi:10.1051/0004-6361/201526750

Afanasiev, A., Vainio, R., Rouillard, A., et al.: Modelling of proton acceleration in application to a ground level enhancement. Astron. Astrophys. (2017, Submitted)

Aran, A., Lario, D., Sanahuja, B., et al.: Astron. Astrophys. **469**, 1123 (2007). doi:10.1051/0004-6361:20077233

Atwood, W.B., Abdo, A.A., Ackermann, M., et al.: Astrophys. J. **697**, 1071 (2009). doi:10.1088/0004-637X/697/2/1071

Battarbee, M.: Acceleration of solar energetic particles in coronal shocks through self-generated turbulence. Ph.D. thesis, University of Turku, Finland (2013). http://urn.fi/URN:ISBN:978-951-29-5574-9

Battarbee, M., Vainio, R., Laitinen, T., Hietala, H.: Astron. Astrophys. **558**, A110 (2013). doi:10.1051/0004-6361/201321348

Jiggens, P.T.A., Gabriel, S.B., Heynderickx, D., et al.: IEEE Trans. Nucl. Sci. **59**, 1066 (2012). doi:110.1109/TNS.2012.2198242

Kocharov, L.G., Torsti, J., Vainio, R., Kovaltsov, G.A.: Sol. Phys. **165**, 205 (1996). doi:10.1007/BF00149100

Kozarev, K.A., Evans, R.M., Schwadron, N.A., et al.: Astrophys. J. **778**, 43 (2013). doi:10.1088/0004-637X/778/1/43

Pomoell, J., Aran, A., Jacobs, C.,et al.: J. Space Weather Space Clim. **5**(27), A12 (2015). doi:10.1051/swsc/2015015

Rouillard, A.P., Plotnikov, I., Pinto, R.F., et al.: Astrophys. J. **833**, 45 (2016). doi:10.3847/1538-4357/833/1/45

Vainio, R.: Astron. Astrophys. **406**, 735 (2003). doi:10.1051/0004-6361:20030822

Chapter 10
Inversion Methodology of Ground Level Enhancements

B. Heber, N. Agueda, R. Bütikofer, D. Galsdorf, K. Herbst, P. Kühl, J. Labrenz, and R. Vainio

Abstract While it is believed that the acceleration of Solar Energetic Particles (SEPs) is powered by the release of magnetic energy at the Sun, the nature, and location of the acceleration are uncertain, i.e. the origin of the highest energy particles is heavily debated. Information about the highest energy SEPs relies on observations by ground-based Neutron Monitors (NMs). SEPs with energies above 500 MeV entering the Earth's atmosphere will lead to an increase of the intensities recorded by NMs on the ground, also known as Ground Level Event or Ground Level Enhancement (GLE). A Fokker-Planck equation well describes the interplanetary transport of near relativistic electrons and protons. An NM is an integral counter defined by its yield function. From the observations of the NM network, the additional solar cosmic ray characteristics (intensity, spectrum, and anisotropy) in the energy range $\gtrsim$500 MeV can be assessed. If the interplanetary magnetic field outside the Earth magnetosphere is known (see Sect. 10.3.2) a computation chain can be set up in order to calculate the count rate increase of an NM for a delta injection at the Sun along the magnetic field line that connects the Sun with the Earth (Sect. 10.3.3). By this computations, we define a set of

---

B. Heber (✉) • D. Galsdorf • K. Herbst • P. Kühl • J. Labrenz
Christian-Albrechts-Universität zu Kiel, Kiel, Germany
e-mail: heber@physik.uni-kiel.de; galsdorf@physik.uni-kiel.de; herbst@physik.uni-kiel.de; kuehl@physik.uni-kiel.de; labrenz@physik.uni-kiel.de

N. Agueda
University of Barcelona, Barcelona, Spain
e-mail: agueda@fqa.ub.edu

R. Bütikofer
University of Bern, Physikalisches Institut, Sidlerstrasse 5, CH-3012 Bern, Switzerland

High Altitude Research Stations Jungfraujoch and Gornergrat, Sidlerstrasse 5, CH-3012 Bern, Switzerland
e-mail: rolf.buetikofer@space.unibe.ch

R. Vainio
Department of Physics and Astronomy, University of Turku, Turku, Finland
e-mail: rami.vainio@utu.fi

O.E. Malandraki, N.B. Crosby (eds.), *Solar Particle Radiation Storms Forecasting and Analysis, The HESPERIA HORIZON 2020 Project and Beyond*, Astrophysics and Space Science Library 444, DOI 10.1007/978-3-319-60051-2_10

Green's functions that can be fitted to an observed GLE to determine the injection time profile. If the latter is compared to remote sensing measurements like radio observations, conclusions of the most probable acceleration process can be drawn.

10.1 Introduction

The Earth is constantly bombarded by high-energy particles, also known as cosmic rays. Those who are not deflected by the geomagnetic field, as discussed in Chap. 5, enter the atmosphere and undergo interactions with atoms and molecules as well as with the nuclei of the atmosphere. Low-energy cosmic rays (<500 MeV) are absorbed in the atmosphere, i.e. no secondary particles can be observed at sea level from primary particles as described in Chap. 6. Most of the primary particles are of galactic origin and are known as Galactic Cosmic Rays (GCR). In solar eruptive events, such as solar flares and Coronal Mass Ejections (CMEs), protons and other ions can be accelerated to high energies (>30 MeV). The acceleration mechanisms are thought to be related to magnetic reconnection in solar flares (Aschwanden 2012) and the shock waves generated by CME (Cliver et al. 2004).

According to the review by Reames (1999) impulsive events are small, electron rich, tend to be enriched in heavy ions and are ^{3}He-rich (^{3}He/^{4}He$\approx$1). They are associated with small magnetic loops in the lower corona with heights smaller than 10^4 km and have ionization states typical of solar flare temperatures, i.e., 10^7 K (Mason 2007). The emission is generated at low altitudes, and fast drift (type III) radio emission, which reflects the electron escape into the interplanetary medium (Klein and Trottet 2001), is observed as well.

Gradual SEP events, in contrast, are believed to originate high in the corona from a CME-driven shock whose extent is consistent with the observations of such SEP over relatively wide longitudinal ranges (Rouillard et al. 2011; Dresing et al. 2014). The association of CMEs with type II emission (see Chaps. 1 and 2 for details) also confirms the conclusion that gradual SEP events are accelerated by large CME-driven shocks (Reames 1999). These events produce some of the highest intensity events observed at Earth at energies up to several GeV (Mewaldt et al. 2012; Kühl et al. 2016).

While this two class picture appears to describe the origin of <50 MeV ions with some success, the origin of SEPs that cause GLEs is less studied due to the lack of detailed observations (Moraal and McCracken 2012). It was found that the GLE spectra tend to be slightly harder than non-GLE spectra and that they are consistent with double power laws (Mewaldt et al. 2012; Kühl et al. 2016). Also, the authors found that the composition of GLEs tends to have higher Fe/O ratios, enrichments in ^{3}He (Wiedenbeck 2013) and highly-ionized charge-states of heavy elements such as Fe. This lead to the conclusion that GLE ions may be accelerated by CME-driven shocks, with quasi-perpendicular shock geometry and the presence of suprathermal ions from previous flares playing a key role (Tylka et al. 2005). Moraal and Caballero-Lopez (2014) found different scenarios for different GLEs having a prompt component from the impulsive phase and a gradual one indicating shock acceleration. To gain further insight the interplanetary and magnetospheric transport of high energy charged particles needs modeling (Bieber et al. 2004). While a Fokker-Planck equation well describes the interplanetary transport of near relativistic electrons (Dröge 2003; Dröge et al. 2010) the processes that need to be included at high energies have not been fully explored (Dalla et al. 2013, 2015).

Gradual particle events tend to be larger, typically associated with large 10^5 km X-ray emitting structures, last much longer than impulsive events, are proton-rich without significant enrichment in ^{3}He/^{4}He, electron poor, and have elemental abundances and charge states representative of solar coronal or solar wind material and temperatures, i.e., 10^6 K (Klecker 2013).

However, the relative roles of both components and how we can discriminate them remains a key problem in solar and solar-terrestrial physics, especially regarding the diverse interest in GLEs. There is a practical interest in GLEs owing to their significance for space weather. Solar cosmic rays can damage spacecraft electronic components and are a significant radiation hazard to astronauts. To quantify these risks, the full particle distribution in energy and pitch angle as a function of time needs to be determined from the NM observations. A method to derive the "physical quantities" is based on forward-modeling of SEP transport (Sects. 10.3.1 and 10.3.2) in interplanetary space and the Earth's magnetosphere (see also Chaps. 4 and 5) by utilizing a power law spectrum in rigidity at the injection point of the Sun (see Chap. 3) and the response/yield function of the NM (see Sect. 10.3.2 and Chap. 6). The forward modeling is utilized in Sect. 10.4 to derive an inversion methodology that is applied to observations in Sect. 10.5. To validate our model chain, results are compared to spacecraft measurements that are described in Sect. 10.2.

10.2 Space and Ground Based Measurements of GLEs

SEP events, where protons are accelerated to energies above 500 MeV, occur a few times per solar cycle. Protons with such energies penetrate the Earth's atmosphere and produce secondary particle showers which can increase the intensities recorded by NMs on the ground. Such intense SEP events are also known as Ground Level Events or GLEs. Initially designed by Simpson (1948), NMs are used for precise monitoring of spectral and directional variations in the cosmic-ray flux. The detection of a GLE event by an NM on average occurs a few times per solar cycle. The first GLE was observed on February 28, 1942 and the first GLE observed by NMs was the one on February 23, 1956 (see gle.oulu.fi).

Since 1942 until the end of 2015 a total of about 70 GLEs have been observed, i.e. $\sim$one GLE per year. Each GLE has its typical characteristics (amplitude, spectrum, duration, spatial distribution of flux, etc.). During a GLE the measurements of the ground-based NM network show an increase in the count rate within typically a few minutes and decreasing intensities to background levels within hours. In some cases, GLEs show a double-peaked time structure, with an initial fast rise and an anisotropic particle population—called "prompt component"—followed by a more gradual and less anisotropic "delayed component" (McCracken et al. 2012).

SEP events causing GLEs are restricted to events that accelerate ions to energies above $\sim$500 MeV/nucleon. Experiments on board spacecraft located close to Earth such as the Geostationary Operational Environmental Satellites (GOES), the Solar and Heliospheric Observatory (SOHO), the Payload for Antimatter Matter Exploration and Light-nuclei Astrophysics (PAMELA) and the Alpha Magnetic Spectrometer 02 (AMS-02) aboard the International Space Station (ISS) can observe protons below and above this threshold in differential energy channels. These data serve to fill an observational gap between the few hundred MeV and less, typically monitored by other satellite instruments on board e.g. the Advanced Composition Explorer (ACE) (Gold et al. 1998) and Wind (Lin et al. 1995). This latter spacecraft allows us to determine the chemical and isotope abundance (Mewaldt et al. 2007; Nitta et al. 2015), the charge states (Mason et al. 1995; Kartavykh et al. 2007) and to study the lepton component i.e. electrons (Dröge 2000; Dresing et al. 2012, 2014; Agueda et al. 2014) during SEP events. As discussed in Chap. 4 it is important to know not only the intensity-time profiles but also the evolution of the PAD. The pitch angle is defined as the angle between the magnetic field and the velocity vector β. A particle instrument with a limited opening angle and a pointing direction $\mathbf{n}$ in interplanetary space is sensitive to a pitch angle range around $\cos(\vartheta) = -\frac{\mathbf{B}\cdot\mathbf{n}}{B}$ with $\mathbf{B}$ the interplanetary magnetic field vector. Spacecraft like ACE, Wind, and the International Monitoring Platform (IMP) 8 outside the Earth's magnetosphere provide us with measurements of the interplanetary magnetic field vector $\mathbf{B}$ near the first Lagrangian point L1. Contrary to IMP 8 neither GOES nor SOHO measures $\mathbf{B}$. The OMNI dataset[1] maintained by the US National Space Science Data Center provides us with the magnetic field, and plasma data sets from ACE, Wind and the IMP-8 shifted to the Earth's bow shock nose. For an observer in the magnetosphere, (PAMELA and AMS-02) the viewing direction $\mathbf{n}$ is the asymptotic direction that has to be calculated as described in Chap. 5.

10.2.1 dE/dx-dE/dx-Method

This method is utilized by EPHIN aboard SOHO and is based on the energy loss in two detectors. Figure 10.1 right illustrates the measurement principle of the EPHIN instrument (Müller-Mellin et al. 1995). Plotting the energy loss in two

[1] http://omniweb.gsfc.nasa.gov/owmin.html.

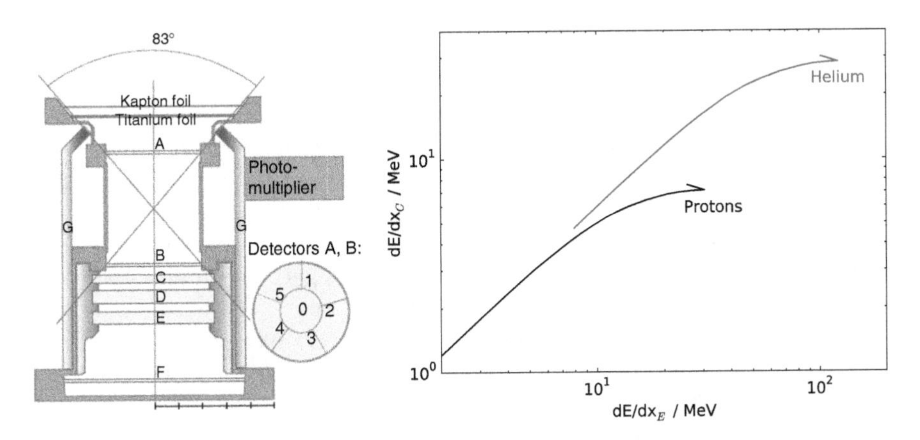

Fig. 10.1 The *left and right panels* display a sketch of the Electron Proton Helium INstrument (EPHIN) aboard SOHO (Müller-Mellin et al. 1995) and the dE/dx-dE/dx measurement principle applied to it, respectively

adjacent detectors against each other, the mean energy losses of H and He follow the characteristic tracks. This is used to identify the particle species in certain areas of the two-dimensional pulse height plane. In these regions, the energy loss in both detectors is attributed to the incoming energy of the particle. Both Goddard Medium Energy (GME) experiment and the Medium Energy Detector (MED) (Meyer and Evenson 1978) utilize this technique.

EPHIN is a multi-element array of solid state detectors with anticoincidence to measure energy spectra of electrons in the range 250 keV to >8.7 MeV, and of hydrogen and helium isotopes in the range 4 MeV/n to >53 MeV/n. The instrument is sketched in Fig. 10.1 left and consists of a stack of silicon solid state detectors (A-F) surrounded by an anticoincidence (scintillator, G). The method to derive energy spectra for penetrating particles with energies above 50 MeV/nucleon is described in detail by Kühl et al. (2015). Since relativistic protons and electrons have the same energy loss dE/dx in matter, electrons with energies above 10 MeV are leading to too high fluxes at energies above $\approx$ 700 MeV for protons when utilizing the $dE/dx - dE/dx$-method.

The second instrument utilizing this method is the GME aboard IMP-8 that was launched by NASA in 1973 into a 35 R_E geocentric orbit with a 12 days period. The spacecraft was in the solar wind for 7–8 days of every 12-day orbit, where it measured the magnetic fields, plasma, and energetic charged particles (e.g., cosmic rays). The spacecraft spin axis was normal to the ecliptic plane, and the spin rate was 23 rpm. PAD information was obtained in eight angular sectors.

In contrast to EPHIN, the MED design consisted of three pulse-height analyzed CsI detectors with thicknesses of 1 mm, 2 cm and 1 mm, respectively, and a cylindrical plastic scintillator anticoincidence shield. Penetrating particles have an energy above 80 MeV and a geometry factor of nearly 5.0 cm^{2}sr (for details see http://spdf.gsfc.nasa.gov/imp8_GME/GME_instrument.html).

10.2.2 dE/dx - C

This method is used for penetrating particles. This method is based on charged particles completely penetrating the Solid State Detector (SSD) and a Cerenkov detector (C), which is placed underneath. If the particles penetrate C faster than with the speed of light in this medium, i.e., with a speed of $\beta > \frac{1}{n}$ where n is the refractive index of the material, they produce a measurable light flash (Cerenkov radiation). Plotting the Cerenkov detector signal against the energy-loss by ionization, ΔE in the SSD results in characteristic curves for protons and helium, clearly separated with their slopes depending on particle speed. Thus, the method allows an identification of the penetrating particles and gives their energy above a threshold speed. This method is utilized by the HEPAD assembly (Rinehart 1978) aboard a series of GOES satellites in geosynchronous orbit maintained by National Oceanic and Atmospheric Administration (NOAA), which provides differential fluxes in three channels between 350 MeV and 700 MeV, and integral flux above 700 MeV (Sauer 1993). HEPAD consists of two 500 μm thick silicon detectors and a silica radiator. The two surface barrier silicon detectors have an area of 3 cm^2 and define an effective acceptance aperture of $\sim$24° half-angle.

Charged particles, which completely penetrate the two semiconductor detectors, can be studied with the help of the Cerenkov-detector (McDonald 1956; Linsley 1955), which is placed underneath the semiconductor detectors. The use of a different detector is necessary since, with energy-loss-measurements alone, the energy determination is only possible in a narrow energy range above the maximum energy for stopping particles. The official archive for GOES Energetic Particle data, including proton flux data, can be found at the National Geophysical Data Center.[2]

10.2.3 Magnet Spectrometer

In addition we utilize data from the Russian-Italian PAMELA mission (Picozza et al. 2007) and the AMS-02 aboard the ISS (Aguilar et al. 2013; Kounine 2012). PAMELA as well as AMS-02 are magnetic spectrometers in Low Earth Orbit, providing extremely high-quality observations of electrons and ions, which we have used for validation and cross-calibration purposes. Because the orbits of PAMELA and AMS are within the Earth's magnetic field, the two spacecraft do not have a 100% duty cycle for observing low-energy cosmic rays (Adriani et al. 2011). Therefore our approach is to use their observations to validate measurements e.g. from EPHIN.

[2]http://www.ngdc.noaa.gov/stp/satellite/goes/.

10.3 Forward Modeling from the Sun to the Observer at Ground

The project HESPERIA gathered experts in different fields to take into account interconnections between the solar, heliospheric and NM communities and to advance our knowledge of GLEs further. The group developed a model chain allowing to infer the solar release time profile of relativistic SEPs and their interplanetary transport parameters directly from NM observations. The process chain starts with the propagation of SEPs from the Sun to the Earth. Utilizing the local interplanetary field in Geocentric Solar Ecliptic (GSE) coordinates, the PAD outside the magnetosphere can be converted to an angular distribution in GSE coordinates. Computing the energetic particle transport in the geomagnetic field and taking into account the NM yield function the count rate variation of all NMs that have measured the event can be predicted. Although the physics of the underlying processes is discussed in detail in Chap. 4 (interplanetary transport), Chap. 5 (transport through the magnetosphere) and Chap. 6 (ground-based measurements by NMs) we will show in a compact way how the different models need to be interlinked.

10.3.1 Interplanetary Particle Transport: From the Sun to the Magnetosphere

Here we give a summary of Chap. 4 in which we model the particle transport from the Sun ($r_i = 2\ r_{Sun}$) in an unperturbed solar wind with constant velocity v. The Interplanetary Magnetic Field (IMF) can be described as a smooth average field, represented by an Archimedean spiral, with superposed magnetic fluctuations. The quantitative treatment of the evolution of the particle's phase space density, $f(z, \mu, t)$, can be described by the focused transport equation (Roelof 1969):

$$\frac{\partial f}{\partial t} + \beta c\, \mu \frac{\partial f}{\partial z} + \frac{1-\mu^2}{2L} \beta c\, \frac{\partial f}{\partial \mu} - \frac{\partial}{\partial \mu}\left(D_{\mu\mu} \frac{\partial f}{\partial \mu} \right) = q(z, \mu, t) \tag{10.1}$$

Here z is the distance along the magnetic field line that depends on the solar wind velocity v, $\mu = \cos\alpha$, is the cosine of the particle pitch angle, α, and t is the time. The systematic force is characterized by the focusing length, $L(z) = -B(z)/(\partial B/\partial z)$, in the diverging magnetic field B, while the stochastic forces are described by the pitch-angle diffusion coefficient $D_{\mu\mu}(\mu)$. As discussed in detail in Chap. 4 the pitch-angle diffusion coefficient has the same form as in Agueda et al. (2008).

Another approximation was introduced by Hasselmann and Wibberenz (1970). If we take the particle radial mean free path, λ_r, to be spatially constant, then the mean free path parallel to the IMF line is given by $\lambda_{||} = \lambda_r \sec^2 \psi$, where ψ is the angle between the field line and the radial direction. As shown in Chap. 4 it is sufficient to compute a database of "Green's-functions" for particles moving with the speed of light ($\beta = 1$). The database consists of 30 different values of the radial mean free path, from $\lambda_r = 0.1$ to $\lambda_r = 2.0$ AU and for solar wind velocities ranging from 300 to 700 km/s. The intensity spectra at the source are given by the solar spectrum that is typically a power-law with $N(R) \propto R^{-\gamma}$, where γ is the spectral index of the solar source. Results of this modeling have been discussed in Chap. 4 and (Agueda 2008). There the results could be directly compared to spacecraft measurements. However, observations in the magnetosphere and ground-based measurements need to take into account the geomagnetic filter as described in Chap. 5 and the shielding by the atmosphere and the specific response of the NM (Chap. 6).

If we assume for each energy interval a **number of delta injections** in time at the source region the phase space density $f(z, \mu, t)$ can be calculated for any time t at 1 AU corresponding to a distance $z(v)$ depending on the solar wind velocity v for every μ if the **mean free path** λ_r, the solar wind speed v and the direction of the interplanetary magnetic field are known. Acceleration theories (see Chap. 3), as well as measurements, suggests that the source spectrum is given by a power law described by the **spectral index** γ. **Thus three free parameters describe the temporal evolution of a SEP event caused by a single δ-injection at the Sun.**

10.3.2 *From the Interplanetary Particle Distribution to Neutron Monitor Measurements - Magneto- and Atmospheric Transport of Charged Energetic Particles*

The transport of cosmic ray particles in the geomagnetic field and in the Earth's atmosphere are described in detail in the Chaps. 5 and 6. Here we refer to the objects that are relevant to the investigations of NM data during solar cosmic ray events. The asymptotic viewing directions for each NM station are often computed only for primary cosmic ray particles that penetrate into the Earth's atmosphere from a vertical direction. The contribution of obliquely incident particles is often neglected in GLE analysis as their contribution to the counting rate is assumed to be small because of the soft spectrum of the solar cosmic rays.

The minimum rigidity that a charged energetic particle must have to reach a location within the magnetosphere and from a given direction of incidence is expressed by the geomagnetic cutoff rigidity R_C. This cutoff rigidity varies from a minimum at the magnetic poles ($R_C \approx 0$ GV) to a value of about 15 GV in equatorial regions. The asymptotic direction for a NM, for a given direction of incidence into the Earth's atmosphere above the location of the NM and a selected particle rigidity, is defined as the direction of motion of the primary cosmic ray particle before penetrating the Earth's magnetosphere. In contrast to previous work by e.g. Bieber et al. (2004), neither the shape of the PAD nor the incoming direction is a free parameter in the High Energy Solar Particle Events forecasting and Analysis (HESPERIA) approach. The PAD of SEPs outside the Earth's magnetosphere is calculated as described in Sect. 10.3.1. To assign for each rigidity the pitch angle range of a NM we need to know the Interplanetary Magnetic Field (IMF) direction close to the Earth magnetopause and the asymptotic direction in GSE coordinates. In HESPERIA, we make use of the 5-min averaged OMNI data to prepare a set of magnetic field directions outside the magnetosphere, to be used as the directions of the symmetry axis of the directional relativistic proton distribution. The field to which the distribution tends to become gyro-tropic is not the momentarily measured field at the particle position but rather the field it averages over. Although the timescales of the gyro motion are relatively small, the spatial ones are not. A first order estimation, $r_L/(u_{SW}t) \sim 1$, leads to $t = 1670$ s for the averaging time of the fluctuating field for a 1-GV particle in a 5 nT field and a solar wind speed of 400 km/s. However, it should be noted that particles would need a rigidity-dependent averaging time of the field direction. As an example, Fig. 10.2 displays on the left the rigidity-dependent vertical asymptotic directions for a stationary observer in Kiel from 1:30 to 3:00 UT on May 17, 2012. The colored triangles show the direction of the interplanetary field for each period. From these two directions, the pitch angle for each rigidity and time can be calculated. The right panel of Fig. 10.2 displays the corresponding results.

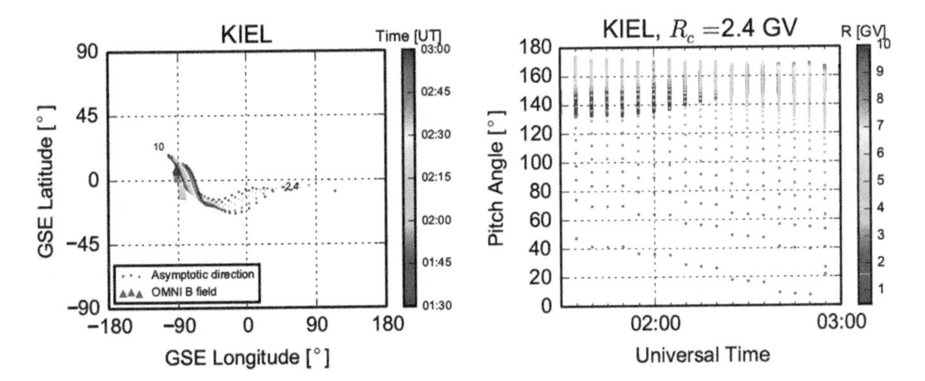

Fig. 10.2 *Left*: Calculations of the asymptotic direction at different rigidities during the onset of the May 17, 2012 SEP event as function of time for an observer at Kiel. The *triangles* indicate the direction of the interplanetary magnetic field. The *right panel* shows the pitch angle coverage for the same period. For details see text

If we assume that only protons are accelerated in a SEP event to sufficient rigidities R that NMs can measure, then the count rate N caused by solar cosmic ray protons at the NM station x and at time t can be expressed by:

$$N_x(t) = \int_{R_c}^{R_u} I(R, \alpha, t) \cdot S_x(R, h) \cdot dR, \tag{10.2}$$

where R_c and R_u are the cutoff and upper ($R_u = 20$ GV) rigidities, α the pitch angle of the incoming proton, $S_x(R, h)$ the yield function of the NM station x at the atmospheric depth, h (for details see Chap. 6). Here Eq. (10.2) is approximated by the sum of discrete values covering the full rigidity and pitch angle range. The upper value of $R_u = 20$ GV was chosen because there is no observational evidence for SEP protons with rigidities above 20 GV.

Given the IMF vector **B** in GSE coordinates and the asymptotic directions **n** in the same coordinate system, we compute the pitch angle, α coverage, for each NM. Together with the yield function described in Chap. 6 the count rate increase for each NM is computed.

10.3.3 Combined Greens-Function

As part of HESPERIA, the two models described in the previous two sections were combined. The first step is to compute the additional count rate, N, caused by SEPs for a given NM station with cutoff rigidity R_C at the time t_1, based on the computed rigidity and cosine of the pitch angle (μ) dependent intensity $I_1(\mu, t_1, r)$ of relativistic protons at time t_1. This spectrum I_1 results from an energy spectrum $I_0 \propto R^{-\gamma}$ injected as a δ-function at a time t_0 before t_1. The two parameters determining the injection function are the power-law index γ and the number of particles injected at a rigidity $R = 1$ GV. The particle transport in the IMF is described by Eq. (10.1) using λ_0 as the only free parameter. The path length of the particles is determined by the solar wind speed measured at 1 AU (see Sect. 10.3.1). The IMF outside the Earth's magnetosphere together with the asymptotic direction calculated for t_1 gives the pitch angle coverage for the NM. Within HESPERIA the Green's function $G_\beta(t, z, \mu, v)$ are derived from the ones $G_c(t, z, \mu, c)$ that are available from SEPServer (http://server.sepserver.eu) for relativistic protons ($\beta = 1$). For near relativistic particles the energy loss in the IMF can be neglected and therefore the Green's function for a lower velocity β is given by:

$$G_\beta\left(\frac{1}{\beta}t, z, \mu, \beta\right) = \beta G_c(t, z, \mu, c)\ . \tag{10.3}$$

We consider near relativistic protons with rigidities from 0.5 to 20 GV (kinetic energies from 0.12 to 19 GeV) and following a power law in rigidity $I(R) = I_0 \cdot R^{-\gamma}$ with the spectral index γ.

The proton mean-free path is taken to increase with rigidity following the standard model. The resulting pitch angle dependent energy spectra at 1 AU are transported through the magnetosphere and atmosphere as described above.

10.4 Inversion Methodology

The previous two sections are describing the forward modeling of energetic particles from the Sun through the inner helio-, magneto- and atmosphere. Up to now, the inversion has only been attempted in two separate steps:

Inversion of NM data: The standard analysis of GLE events is based on solving an inverse problem where data by the worldwide network of NMs is used to determine the spectral and angular characteristics of SEPs near Earth but outside the magnetosphere causing the GLE.

Inversion of spacecraft data: The injection time profile of SEPs at the Sun and the characteristics (mean free path) of their transport in the interplanetary medium is inverted from 1 AU spacecraft measurements.

In what follows we review first the two inversion approaches used so far and then describe the HESPERIA approach.

10.4.1 Inversion of Spacecraft Data to the Sun

Numerical simulations of the propagation of SEPs along the IMF are a useful tool to understand SEP events and their sources. We currently have a good theoretical understanding of the transport processes that affect SEPs in the interplanetary medium (see Chap. 4). In Sect. 10.3.1 we discussed a model that simulates the processes undergone by SEPs during their propagation from their source to the observer with the critical parameters summarized in the gray box on page 186.

The approach introduced by Agueda et al. (2008) is to utilize the computed response of the "system" to an impulsive (delta) injection at the Sun, i.e. the Green's function of particle transport. A convolution of some delta injections allows us to compute different pitch angle dependent proton intensity time profiles that are used as input to the second step in the chain (see Sect. 10.3.2).

Given a system impulse response, $g(t)$, and the input injection profile $q(t)$, the output, $I(t)$, is the convolution of $g(t)$ and $q(t)$:

$$I(t) = \int_0^\infty g(t')q(t - t')dt' \qquad (10.4)$$

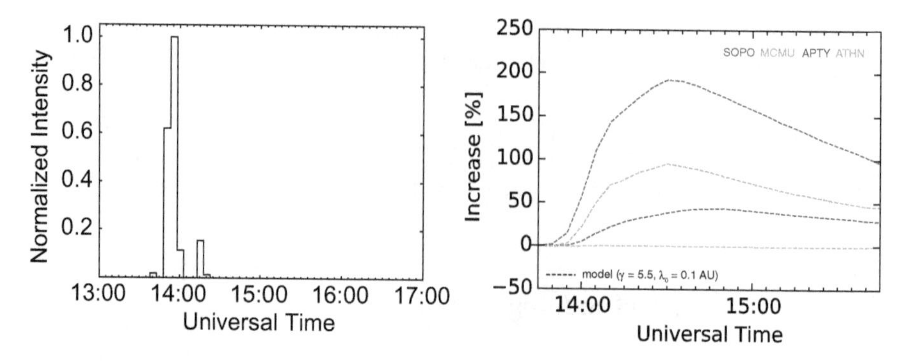

Fig. 10.3 Convolution of different δ-injections close to the Sun on April 15, 2001 (GLE60) from 13:45 h to 14:15 h (*left panel*) with a set of δ injection $\gamma = 5.5$ at the Sun, and a mean free path $\lambda = 0.1$ AU in interplanetary space leads to intensity increases at the Apatity, Athens, McMurdo and South Pole NM (*right panel*)

i.e., $I(t)$ is the sum of responses resulting from a series of impulses at the Sun, weighted and shifted in time according to $q(t)$. For simplicity, here we assumed that the response is only time-dependent, but the same holds if one includes the energetic and directional dependence in the Green's function.

Particle intensities measured in the heliosphere as a function of time, energy and direction, are obtained as a temporal convolution of the source function (particle injection profile) and the Green's function of particle transport at the spacecraft location. The origin of SEP events can be unfolded by solving the inverse problem (deconvolving the in-situ measurements). The measurements are used to infer the values of the model parameters. It is a deductive approach, and it has the advantage that a systematic exploration of the parameters' space is possible. A simulated GLE time profile with a proton injection from 13:45 h to 14:15 h at the Sun as displayed in the left panel of Fig 10.3 leads to a proton event observable close to Earth by intensity increases at Apatity, Athens, McMurdo and the South Pole displayed in the right panel. A convolution of a δ-injection with $\gamma = 5.5$, and a mean free path $\lambda_0 = 0.1$ AU has been used.

Let's consider an arbitrary function $q(t)$—to be determined—that represents the injection function of SEPs close to the Sun. The modeled directional intensities, M_j^k, resulting from a series of impulse solar injections can be written as

$$M_j^k(t;\lambda) = \int_{T_1}^{T_2} dt'\, g_j^k(t,t';\lambda)\, q(t'), \tag{10.5}$$

where $g_j^k(t,t';\lambda)$ represents the impulse response in a given direction j and energy interval k, at a given time t, when particle injection took place at time t' and assuming an interplanetary mean free path λ. The duration of the injection function, $t' \in [T_1, T_2]$, is determined by the SEP event time interval selected for fitting, $t \in [t_1, t_2]$, that is, $T_1 = t_1 - \Delta t$ and $T_2 = t_2 - \Delta t$, where Δt is the transit time of the first arriving particles at the spacecraft location, assuming a given value of the scattering mean free path. The number of time points in the event time interval selected for fitting is equal to $n_t = (t_2 - t_1)/\delta t + 2$, where δt is the time resolution of the data.

Taking discrete values of time, we have

$$M_j^k(t_h; \lambda) = \sum_{l=1}^{n_t} g_j^k(t_h, t_l'; \lambda)\, q(t_l') \tag{10.6}$$

where $j = 1, 2, \ldots, n_s$ are numbers representing the directions (sectors or bins) observed by the telescope, $k = 1, 2, \ldots, n_c$ numbers for the energy channels and $h = 1, 2, \ldots, n_t$ numbers of the time intervals.

Equation (10.6) can be written as

$$M_j^k(t_h; \lambda) \equiv M_i(\lambda) = \sum_{l=1}^{n_t} g_{il}(\lambda)\, q_l = (\mathbf{g} \cdot \mathbf{q})_i \tag{10.7}$$

where $i = k + (j-1) \cdot n_t = 1, 2, \ldots, n_T$ numbers the total number of observational points and $n_T = n_t\, n_s$ gives the total number of observational points in all sectors; $\mathbf{g}$ is an $n_T \times n_t$ matrix with $(\mathbf{g})_{il} = g_{il}(\lambda)$.

In matrix form,

$$\begin{pmatrix} M_1 \\ M_2 \\ \vdots \\ M_{n_T} \end{pmatrix} = \begin{pmatrix} g_{11} & g_{12} & \cdots & g_{1n_t} \\ g_{21} & g_{22} & \cdots & g_{2n_t} \\ \vdots & \vdots & \ddots & \vdots \\ g_{n_T1} & g_{n_T2} & \cdots & g_{n_Tn_t} \end{pmatrix} \cdot \begin{pmatrix} q_1 \\ q_2 \\ \vdots \\ q_{n_t} \end{pmatrix} \tag{10.8}$$

The goal is to compare the modeled intensities with the observations. Let J_i be the observations (background subtracted). We want to derive the n_t-vector $\mathbf{q}$ that minimizes the length of the vector $\mathbf{J} - \mathbf{M}$, that means minimizing the value of

$$||\mathbf{J} - \mathbf{M}|| \equiv ||\mathbf{J} - \mathbf{g} \cdot \mathbf{q}||, \tag{10.9}$$

subject to the constraint that $q_l \geq 0 \;\; \forall l = 1, 2, \ldots, n_t$. Thus, the best-fit $\mathbf{q} = (q_1, q_2, \ldots, q_{n_t})$ corresponds to a combination of delta-function injection amplitudes. To obtain the best-fit values, we use the NNLS method developed by Lawson and Hanson (1974), which always converges to a solution.

Note, that if each energy channel is fitted separately, the total number of observational points considered in the fit is $n_t n_s$ and the total number of independent fitting parameters (injection amplitudes) is n_t. Since $n_t n_s \gg n_t$, the number of degrees of freedom is clearly much larger than the number of model parameters and the inversion problem is well constrained. If one instead neglects the directional information in the data and uses the modeled omnidirectional intensities to fit the omnidirectional event, then the number of observational points and the number of independent fitting parameters is equal, and the problem is not well constrained (multiple injections and transport scenarios can provide an explanation for the data).

The goodness of fit describes how well the model predictions fit a set of observations. One way to evaluate the goodness of the fit, in case the measurement errors are known, is to construct a weighted sum of squared residuals (for details see Agueda 2008). The χ^2 estimator does not work very well for SEP events because during impulsive events the maximum intensities can be several orders of magnitude higher than the intensities observed during the decay phase, thus emphasizing the peak period. Therefore a better goodness-of-fit estimator is provided by the sum of the squared logarithmic differences between the observational and the modeled data. This estimator gives an equal weight of all relative residuals instead of just emphasizing the goodness of fit at the time of maximum. By evaluating the goodness of the fit under different interplanetary transport conditions (different values of λ), one can objectively discern the "best fit" scenario (λ-value and associated injection profile) by minimizing the values of the goodness-of-fit estimator.

10.4.2 Inversion of NM Data to the Border of the Earth's Magnetosphere

In general, the standard analysis of GLE events is based on an inverse problem, where data by the worldwide network of NMs is used to determine the spectral and angular characteristics of SEPs near Earth causing GLEs for selected times (Shea and Smart 1982; Mishev et al. 2014). Analysis of the characteristics of the primary solar particles causing GLEs from ground-based data records is a serious challenge (Bütikofer and Flückiger 2013). Data from stations with different cutoff rigidities (geomagnetic latitudes) provide information necessary to determine the spectral characteristics. Responses of stations over a wide range of geographical locations are required to determine the axis of symmetry. Data are fitted to directional distributions that are rotationally symmetric about one direction in space. This, in principle, is close but not exactly the direction of the instantaneous magnetic field measured close to the Earth but outside the Earth's magnetosphere (Bieber et al. 2013). Therefore, axis-symmetry is assumed, but the direction of the axis of symmetry is optimized to fit the data.

The PADs of relativistic solar protons in space are assumed to follow a given functionality. A variety of functions has been used in the literature. These include a linear form, an exponential plus a constant, a parabola, two Gaussians, and two exponentials plus constant. The latter three are expected to provide better fits to bidirectional fluxes, if present (Bieber et al. 2013; Mishev et al. 2014).

Similarly, the spectra of relativistic solar protons during a GLE are assumed to follow a power law in rigidity, or energy with extensions that describe the softening of the spectrum at higher energies by multiplying a power law with an exponential cutoff.

The parameters of the rigidity spectrum, the PAD and the direction of symmetry are determined by minimizing the sum of squared differences between the modeled relative change in the NM count rate at the station x
$\left(\frac{\Delta N_x}{N_x}\right)_{\text{mod.}}$ and
the corresponding observed relative count rate change
$\left(\frac{\Delta N_x}{N_x}\right)_{\text{obs.}}$

$$F = \sum_{x=1}^{m} \left[\left(\frac{\Delta N_x}{N_x}\right)_{\text{mod.}} - \left(\frac{\Delta N_x}{N_x}\right)_{\text{obs.}} \right]^2 \tag{10.10}$$

for the set of selected NM data. The Levenberg-Marquardt algorithm (LMA) (Marquardt 1963) provides a numerical solution to the problem of minimizing a nonlinear function over the space of parameters of the function. The goodness of the fit is commonly expressed by a weighted sum of squared residuals or by computing a correlation coefficient, ρ between measurements and the model.

10.4.3 The HESPERIA Approach

For the first time, models of the transport of SEPs in the interplanetary medium, the Earth's magnetosphere and atmosphere, and the response of NMs are linked to each other. The first goal is to compute the expected additional count rate, N, caused by SEP for a given NM station, based on the intensities of the primary cosmic rays near Earth, $I(R, \alpha, t)$, as function of rigidity, R, pitch angle, α, and time, t. As detailed above the interplanetary transport is described by Green's functions representing characteristics of the SEP interplanetary transport conditions. To ascribe the magnetospheric transport, the magnetic field direction outside the magnetosphere is computed from interplanetary measurements, and the asymptotic directions are calculated utilizing the PLANETOCOSMICS code in GSE coordinates. The latter can be found at the HESPERIA webpage during each GLE in the past and for the NMs of the worldwide network.

Since the parameters are the same than the ones used by Agueda et al. (2008) the differences lie in the indirect measurement of the pitch angle dependent count rate and the integration over wide energy ranges: For all NM stations, the counting rate increases can then be computed for a series of delta injections from the Sun and for selected times. Note that the hardness of the source spectra described by γ is assumed to be constant in the HESPERIA approach for all δ-injections. The amplitudes of the source components, for a given scenario, are inferred by fitting the NM observations with the modeled NM counting rate increases. The NNLS algorithm described in Sect. 10.4.1 is used to determine the best set of parameters. Regularized inversion approaches will be explored for refinements. The goodness of the fit will be evaluated by computing a weighted sum of the squared residuals.

The result of the inversion problem is a detailed time profile of the injection process at the Sun. The shape of this profile is presumably determined by details of the acceleration process and possible transport processes in the corona (see Chap. 3 for details).

In contrast to a classical approach with a total number of at least 10 fit parameters to derive the injection function at the Sun (see Sect. 10.4.2) the HESPERIA approach relies on several well-documented assumptions (see Chaps. 3, 4, and 5) reducing the number of free parameters. Assuming for each rigidity a number of delta injection at the source region the phase space density $f(z, \mu, t)$ close to Earth can be calculated for any given time t corresponding to the distance $z(v)$ that depends on the measured solar wind velocity v. The mean free path λ_0, the number of particles, the injection time profile and its rigidity spectrum described by a power law with index γ are utilized to derive $f(z, \mu, t)$. The cosmic ray particle trajectories through the Earth's magnetosphere are computed for selected rigidities R_c, $R_c + 0.1$ GV ... 20 GV as a function of GSE coordinates of the NM station, for a selected time. Applying the Tsyganenko 1989 (Tsy89) model and the yield function for each NM we can ascribe the SEP time profiles for the NM network by a minimum of free parameters.

10.5 Results and Validation

Since the HESPERIA approach depends on the knowledge of the IMF outside the Earth's magnetosphere that is provided by OMNI since November 4, 1973, our investigation is restricted to GLEs with numbers larger than number 26 occurring on April 30, 1976. To validate the results, pitch angle dependent intensity time profiles of protons with energies above 500 MeV are needed. Here we utilize SOHO/EPHIN and Wind 3DP measurements. The latter are needed in order to estimate the pitch angle coverage of SOHO/EPHIN.

Since SOHO was launched in December 1995, we restrict our validation to GLEs with numbers $\geq$55 (GLE55 occurred on November 6, 1997). Kühl et al. (2015, 2016) showed that EPHIN is capable of measuring the proton spectra in the range from 100 MeV to above 1 GeV. Since no magnetic field measurements are available on SOHO, we compare the intensity-time-profiles of near relativistic electrons measured by EPHIN and Wind 3DP with each other. In Fig. 10.4 different colored curves show the pitch angle dependent time-profiles for Wind 3DP with energies between 230 and 646 keV. Three thick black lines show the EPHIN measurements with energies between 0.25 and 0.7 MeV multiplied by different factors (0.05, 0.08, and 0.1 as solid, dashed, and dashed-dotted lines) to take into account cross

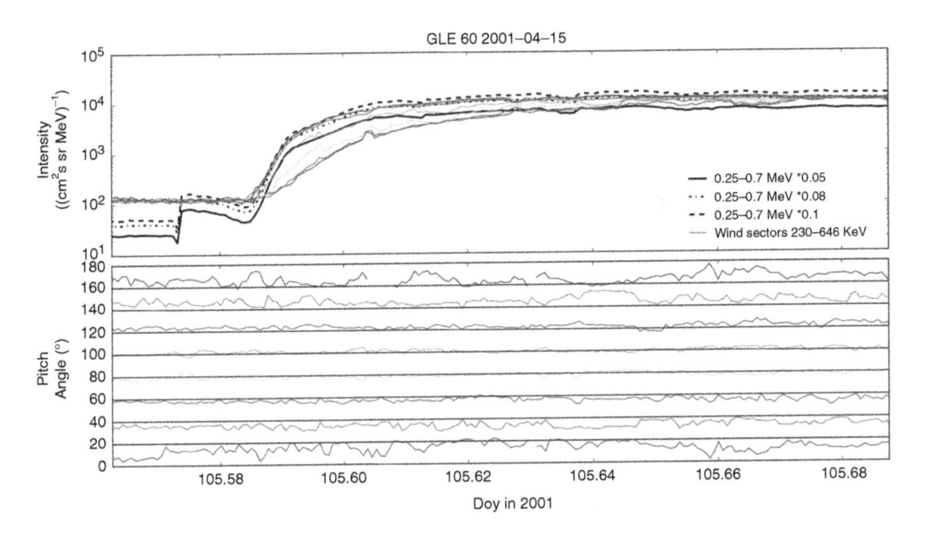

Fig. 10.4 *From top to bottom:* Hourly averaged intensity time profiles of 234–646 keV electrons measured by Wind 3DP. The different colors give the pitch angle coverage for the WIND observations shown in the *bottom panels*. Also the *solid, dashed and dashed-dotted lines* represent the 0.25–0.7 MeV EPHIN electron measurements multiplied by a factor 0.05, 0.08, and 0.1 respectively

calibration. Only the dashed-dotted line agrees with all Wind sectors when the flux becomes isotropic. Thus we assume that the curve that agrees best with the dashed-dotted line ascribes the viewing direction of EPHIN. In our example, the purple curve describing the measurements at a pitch angle of 120° fulfills this criteria best.

Figure 10.5 is an extension of Fig. 10.3 that only showed predicted intensity time profiles close to Earth and by different NMs. It compares the results computed with the GLE inversion software with measurements of other facilities as radio telescopes, NMs and particle detectors in space. The parameters used in the prediction were derived by the inversion of the NM data for GLE 60. The middle panel of Fig. 10.5 compares the simulated and measured intensity time profiles of the Apatity, Athens, McMurdo and South Pole NMs with each other showing a good agreement between measurements and the model. The injection profile close to the Sun is shown in the left panel together with the micro wave profile measured by the Radio Solar Telescope Network (RSTN) showing a good agreement between the particle component injected into interplanetary space and towards the lower corona. Note, that the injection profile is in very good agreement with the one derived by Bieber et al. (2004). For details on micro waves and their importance, the reader is referred to Chap. 2.

The middle bottom panel displays the energy spectrum between 100 MeV and 1 GeV predicted by the model (red curve) and measured by EPHIN (black curve). Note, that the prediction was scaled down by a factor of $\sim$20. From the figure, it is evident that the model predicts the same spectral index when taking into account the contribution of electrons at 900 and 700 MeV, respectively (for details see Kühl et al. 2017).

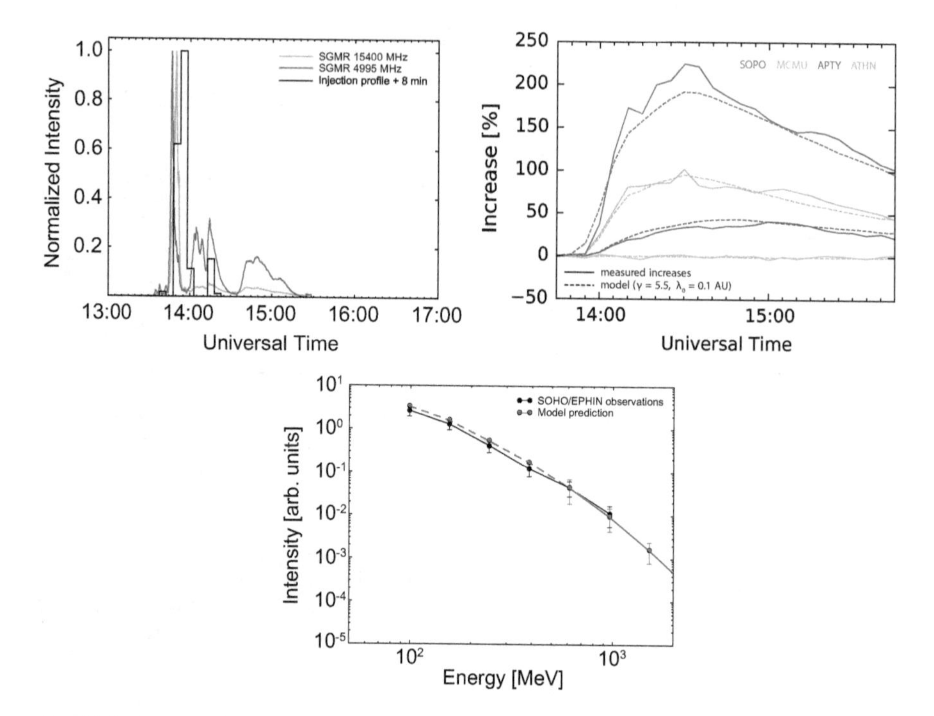

Fig. 10.5 Predicted δ-injections (*upper left panel*), NM increases (*upper right panel*) and differential intensity spectrum (*lower panel*) for GLE60 (see also Fig. 10.3) in comparison to actual measurements (for details see text)

10.6 Concluding Remarks

A new approach has been presented here that allows computing the injection function of SEP close to the Sun based on the data of the worldwide NM network during a GLE. This injection function is described by a power law in rigidity R with two parameters that are the intensity at $R_0 = 1$ GV and the spectral index γ. For the interplanetary transport, we utilize a 1-dimensional model with the mean free path λ_r as free parameter. The solar wind speed taken from interplanetary measurements determines the length of the magnetic field line connecting the Earth back to the Sun. The IMF direction is taken from the OMNI data set using appropriate accumulation periods. The transport through the Earth's magnetosphere is computed with the PLANETOCOSMICS software using the Tsy89 model for the outer Earth's magnetic field. The NM yield function describes the passage of the cosmic ray particles through the atmosphere and the detection of the secondary nucleon component by the NMs. Using this model chain GLE 60 could be reproduced very well. A discrepancy between the prediction and the measurements in interplanetary space exists that needs further investigations. Especially the whole number of observed GLEs should be analyzed and validated by using GOES HEPAD data that go back to the 1970s. Further improvements like utilizing more sophisticated models of the magnetosphere and the interplanetary transport, as well as different NM yield functions, should be taken into account in order to improve the model presented here.

References

Adriani, O., Barbarino, G.C., Bazilevskaya, G.A., Bellotti, R., Boezio, M., Bogomolov, E.A., Bonechi, L., Bongi, M., Bonvicini, V., Borisov, S., et al.: Astrophys. J. **742**, 102 (2011). doi:10.1088/0004-637X/742/2/102

Agueda, N.: Near-relativistic electron events. Monte Carlo simulations of solar injection and interplanetary transport. Ph.D. thesis, Dep. Astronomia i Meteorologia University of Barcelona Martí i Franquès 1 08028 Barcelona (2008)

Agueda, N., Vainio, R., Lario, D., Sanahuja, B.: Astrophys. J. **675**, 1601 (2008). doi:10.1086/527527

Aguilar, M., Alberti, G., Alpat, B., Alvino, A., Ambrosi, G., Andeen, K., Anderhub, H., Arruda, L., Azzarello, P., Bachlechner, A., et al.: Phys. Rev. Lett. **110**(14), 141102 (2013). doi:10.1103/PhysRevLett.110.141102

Agueda, N., Klein, K.L., Vilmer, N., Rodríguez-Gasén, R., Malandraki, O.E., Papaioannou, A., Subirà, M., Sanahuja, B., Valtonen, E., Dröge, W., et al.: Astron. Astrophys. **570**, A5 (2014). doi:10.1051/0004-6361/201423549

Aschwanden, M.J.: Space Sci. Rev. **171**, 3 (2012). doi:10.1007/s11214-011-9865-x

Bieber, J.W., Evenson, P., Dröge, W., Pyle, R., Ruffolo, D., Rujiwarodom, M., Tooprakai, P., Khumlumlert, T.: Astrophys. J. Lett. **601**, L103 (2004). doi:10.1086/381801

Bieber, J.W., Clem, J., Evenson, P., Pyle, R., Sáiz, A., Ruffolo, D.: Astrophys. J. **771**, 92 (2013). doi:10.1088/0004-637X/771/2/92

Bütikofer, R., Flückiger, E.O.: J. Phys. Conf. Ser. **409**(1), 012166 (2013). doi:10.1088/1742-6596/409/1/012166

Cliver, E.W., Kahler, S.W., Reames, D.V.: Astrophys. J. **605**, 902 (2004). doi:10.1086/382651

Dalla, S., Marsh, M.S., Kelly, J., Laitinen, T.: J. Geophys. Res. (Space Phys.) **118**, 5979 (2013). doi:10.1002/jgra.50589

Dalla, S., Marsh, M.S., Laitinen, T.: Astrophys. J. **808**, 62 (2015). doi:10.1088/0004-637X/808/1/62

Dresing, R., Gómez-Herrero, N., Klassen, A., Heber, B., Kartavykh, Y., Dröge, W.: Sol. Phys. **281**, 281 (2012). doi:10.1007/s11207-012-0049-y

Dresing, N., Gómez-Herrero, R., Heber, B., Klassen, A., Malandraki, O., Dröge, W., Kartavykh, Y.: Astron. Astrophys. **567**, A27 (2014). doi:10.1051/0004-6361/201423789

Dröge, W.: Astrophys. J. **537**, 1073 (2000). doi:10.1086/309080

Dröge, W.: Astrophys. J. **589**, 1027 (2003). doi:10.1086/374812

Dröge, W., Kartavykh, Y.Y., Klecker, B., Kovaltsov, G.A.: Astrophys. J. **709**, 912 (2010). doi:10.1088/0004-637X/709/2/912

Gold, R.E., Krimigis, S.M., Hawkins, S.E.I., Haggerty, D.K., Lohr, D.A., Fiore, E., Armstrong, T.P., Holland, G., Lanzerotti, L.J.: Space Sci. Rev. **86**(1), 541 (1998)

Hasselmann, K., Wibberenz, G.: Astrophys. J. **162**, 1049 (1970)

Kartavykh, Y.Y., Dröge, W., Klecker, B., Mason, G.M., Möbius, E., Popecki, M., Krucker, S.: Astrophys. J. **671**, 947 (2007). doi:10.1086/522687

Klecker, B.: J. Phys. Conf. Ser. **409**(1), 012015 (2013). doi:10.1088/1742-6596/409/1/012015
Klein, K.L., Trottet, G.: Space Sci. Rev. **95**, 215 (2001)
Kounine, A.: Int. J. Modern Phys. E **21**(08), 1230005 (2012). doi:10.1142/S0218301312300056. http://www.worldscientific.com/doi/abs/10.1142/S0218301312300056
Kühl, P., Banjac, S., Dresing, N., Goméz-Herrero, R., Heber, B., Klassen, A., Terasa, C.: Astron. Astrophys. **576**, A120 (2015). doi:10.1051/0004-6361/201424874
Kühl, P., Dresing, N., Heber, B., Klassen, A.: Solar energetic particle events with protons above 500 MeV between 1995 and 2015 measured with SOHO/EPHIN. Sol. Phys. **292**(10), 10 (2017). arXiv:1611.03289. doi:10.1007/s11207-016-1033-8. http://adsabs.harvard.edu/abs/2017SoPh..292...10K
Kühl, P., Dresing, N., Heber, B., Klassen, A.: Sol. Phys. **292**, 10 (2017). doi:10.1007/s11207-016-1033-8
Lawson, C.L., Hanson, R.J.: Solving Least Squares Problems. SIAM, Philadelphia (1974)
Lin, R.P., Anderson, K.A., Ashford, S., Carlson, C., Curtis, D., Ergun, R., Larson, D., McFadden, J., McCarthy, M., Parks, G.K., et al.: Space Sci. Rev. **71**, 125 (1995). doi:10.1007/BF00751328
Linsley, J.: Phys. Rev. **97**, 1292 (1955). doi:10.1103/PhysRev.97.1292
Marquardt, D.W.: J. Soc. Ind. Appl. Math. **11**(2), 431 (1963). doi:10.1137/0111030
Mason, G.M.: Space Sci. Rev. **130**, 231 (2007). doi:10.1007/s11214-007-9156-8
Mason, G.M., Mazur, J.E., Looper, M.D., Mewaldt, R.A.: Astrophys. J. **452**, 901 (1995). doi:10.1086/176358
McCracken, K.G., Moraal, H., Shea, M.A.: Astrophys. J. **761**, 101 (2012). doi:10.1088/0004-637X/761/2/101
McDonald, F.B.: Phys. Rev. **104**, 1723 (1956). doi:10.1103/PhysRev.104.1723
Mewaldt, R.A., Cohen, C.M.S., Mason, G.M., Cummings, A.C., Desai, M.I., Leske, R.A., Raines, J., Stone, E.C., Wiedenbeck, M.E., von Rosenvinge, T.T., et al.: Space Sci. Rev. **130**, 207 (2007). doi:10.1007/s11214-007-9187-1
Mewaldt, R.A., Looper, M.D., Cohen, C.M.S., Haggerty, D.K., Labrador, A.W., Leske, R.A., Mason, G.M., Mazur, J.E., von Rosenvinge, T.T.: Space Sci. Rev. **171**, 97 (2012). doi:10.1007/s11214-012-9884-2
Meyer, P., Evenson, P.: IEEE Trans. Geosci. Electron. **16**, 180 (1978)
Mishev, A.L., Kocharov, L.G., Usoskin, I.G.: J. Geophys. Res. **119**, 670 (2014). doi:10.1002/2013JA019253
Moraal, H., Caballero-Lopez, R.A.: Astrophys. J. **790**, 154 (2014). doi:10.1088/0004-637X/790/2/154
Moraal, H., McCracken, K.G.: Space Sci. Rev. **171**, 85 (2012). doi:10.1007/s11214-011-9742-7
Müller-Mellin, R., Kunow, H., Fleißner, V., Pehlke, E., Rode, E., Röschmann, N., Scharmberg, C., Sierks, H., Rusznyak, P., McKenna-Lawlor, S., et al.: Sol. Phys. **162**(1), 483 (1995)
Nitta, N.V., Mason, G.M., Wang, L., Cohen, C.M.S., Wiedenbeck, M.E.: Astrophys. J. **806**, 235 (2015). doi:10.1088/0004-637X/806/2/235
Picozza, P., Galper, A.M., Castellini, G., Adriani, O., Altamura, F., Ambriola, M., Barbarino, G.C., Basili, A., Bazilevskaja, G.A., Bencardino, R., et al.: arXiv.org (4), 296 (2006)
Reames, D.V.: Space Sci. Rev. **90**, 413 (1999). doi:10.1023/A:1005105831781
Rinehart, M.C.: Nucl. Inst. Methods **154**, 303 (1978). doi:10.1016/0029-554X(78)90414-7
Roelof, E.C.: Propagation of solar cosmic rays in the interplanetary magnetic field. In: Ögelman, H., Wayland, J.R. (Eds.) Lectures in High-Energy Astrophysics, p. 111 (1969). http://adsabs.harvard.edu/abs/1969lhea.conf..111R. Provided by the SAO/NASA Astrophysics Data System
Rouillard, A.P., Odstrecil, D., Sheeley, N.R., Tylka, A., Vourlidas, A., Mason, G., Wu, C.C., Savani, N.P., Wood, B.E., Ng, C.K., et al.: Astrophys. J. **735**, 7 (2011). doi:10.1088/0004-637X/735/1/7
Sauer, H.H.: Int. Cosmic Ray Conf. **3**, 250 (1993)
Shea, M.A., Smart, D.F.: Space Sci. Rev. **32**, 251 (1982). doi:10.1007/BF00225188

Simpson, J.A.: Phys. Rev. **73**, 1389 (1948). doi:10.1103/PhysRev.73.1389
Tylka, A.J., Cohen, C.M.S., Dietrich, W.F., Lee, M.A., Maclennan, C.G., Mewaldt, R.A., Ng, C.K., Reames, D.V.: Astrophys. J. **625**, 474 (2005). doi:10.1086/429384
Wiedenbeck, M.E., Mason, G.M., Cohen, C.M.S., Nitta, N.V., Gómez-Herrero, R., Haggerty, D.K.: Astrophys. J. **762**, 54 (2013). doi:10.1088/0004-637X/762/1/54

Index

O.E. Malandraki, N.B. Crosby (eds.), *Solar Particle Radiation Storms Forecasting and Analysis, The HESPERIA HORIZON 2020 Project and Beyond*, Astrophysics and Space Science Library 444, DOI 10.1007/978-3-319-60051-2